Oxford International Primary Maths

Workbook

6

Tony Cotton

Great Clarendon Street, Oxford, OX2 6DP, United Kingdom

Oxford University Press is a department of the University of Oxford. It furthers the University's objective of excellence in research, scholarship, and education by publishing worldwide. Oxford is a registered trade mark of Oxford University Press in the UK and in certain other countries

© Tony Cotton 2015

The moral rights of the author have been asserted

First published in 2015

All rights reserved. No part of this publication may be reproduced, stored in a retrieval system, or transmitted, in any form or by any means, without the prior permission in writing of Oxford University Press, or as expressly permitted by law, by licence or under terms agreed with the appropriate reprographics rights organization. Enquiries concerning reproduction outside the scope of the above should be sent to the Rights Department, Oxford University Press, at the address above.

You must not circulate this work in any other form and you must impose this same condition on any acquirer

British Library Cataloguing in Publication Data
Data available

978 0 19 836531 0

3 5 7 9 10 8 6 4 2

Paper used in the production of this book is a natural, recyclable product made from wood grown in sustainable forests. The manufacturing process conforms to the environmental regulations of the country of origin.

Printed in Great Britain

Acknowledgements

The publishers would like to thank the following for permissions to use their photographs:

Cover: Patryk Kosmider/Shutterstock; **p59l-r:** Arena Photo UK/Shutterstock; Mega Pixel/Shutterstock; Bojan Pavlukovic/Shutterstock; PeterVrabel/Shutterstock; urbanbuzz/Shutterstock; **p74l-r:** Pter Malyshev/Shutterstock; Kitch Bain/Shutterstock; Vladislav Lyutov/Shutterstock; **p86:** Destinyweddingstudio/Shutterstock; **p95:** eurobanks/Shutterstock;

Illustrations: OKS Group

Although we have made every effort to trace and contact all copyright holders before publication this has not been possible in all cases. If notified, the publisher will rectify any errors or omissions at the earliest opportunity.

Links to third party websites are provided by Oxford in good faith and for information only. Oxford disclaims any responsibility for the materials contained in any third party website referenced in this work.

Contents

Unit 1 Number and Place Value
1A/B	The number system and place value	6
1C	Number properties	8
1D/E	Comparing numbers and estimation and rounding	10
1F	Number sequences	12
1	Review	14

Unit 2 Fractions and Decimals
2A	Equivalent fractions	16
2B	Fractions and decimals	18
2C	Addition pairs	20
2D	Mixed numbers and improper fractions	22
2E	Ratio and proportion	24
2F	Percentages	26
2	Review	28

Unit 3 Mental Calculation
3A/F	Mental strategies for addition and subtraction	30
3B/E	Mental strategies for multiplication and division	32
3C	Using known facts to derive new ones	34
3D	Doubling and halving	36
3	Review	38

Unit 4 Addition and Subtraction
4A	Adding and subtracting 3-digit numbers	40
4B	Adding and subtracting money	42
4C	Using negative numbers	44
4	Review	46

Unit 5 Multiplication and Division
5A	Multiplying by 2-, 3- and 4-digit numbers	48
5B	Dividing 3-digit numbers by 2-digit numbers	50
5C	Division with remainders	52
5D	Using the arithmetical laws for multiplication and division	54
5	Review	56

Unit 6 Shapes and Geometry

6A	Classifying polygons	58
6B	Properties of 3D shapes	59
6C/D	Making 2D representations of 3D shapes and angles in a triangle	60
6	Review	62

Unit 7 Position and Movement

7A	Reading and plotting coordinates	64
7B	Reflections and rotations	66
7	Review	68

Unit 8 Length, Mass and Capacity

8A	Selecting and using appropriate units of measure	70
8B	Converting units of measure	72
8C	Using scales and measuring accurately	74
8	Review	76

Unit 9 Time

9A	Converting between units of time	78
9B	Using the 24-hour clock and timetables	80
9C	Calculating time intervals including time zones	82
9	Review	84

Unit 10 Area and Perimeter

10A	Area and perimeter of rectilinear shapes	86
10B	Estimating areas of irregular shapes by counting squares	88
10C	Calculating areas and perimeters of compound shapes	90
10	Review	92

Unit 11 Handling Data

11A	Handling data	94
11B	Probability	95

1 Number and Place Value

Introduction

This unit develops students' understanding of place value and number sequences, both of which have been introduced in earlier stages. In the decimal number system the value of a digit depends on its place, or position in a number. For example, the number 5 642 537 is five million six hundred and forty-two thousand five hundred and thirty-seven. The 5 at the beginning of the number is worth 5 million and the 2 is worth 2 thousands.

To understand number sequences, students need to have clear mental images. Hundred squares and number lines will help students to develop these mental images, and they can use blank number lines to help them build their own number sequences. It is important to continue using these visual representations of the number system at Stage 6.

Ways to help

Many people say that when we multiply by 10 we 'add a zero' and that when we multiply by 100 we 'add two zeros'. This does work with whole numbers but leads to mistakes when we are working with decimal numbers. For example, the answer to 24.31 × 100 is not 24.3100, but 2431. A better strategy is to imagine a place-value grid and move the digits two places to the left:

	Thousands	Hundreds	Tens	Units	.	Tenths	Hundredths
			2	4	.	3	1
× 100	2	4	3	1	.		

There are activities in this unit which involve counting on and counting back across zero. Encourage students to use a number line so that they can see how this works. For example: 25 − 75 = −50.

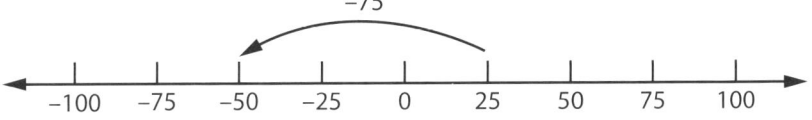

Encourage students to say the complete calculation: 'Twenty-five minus seventy-five equals negative fifty.'

Key Words

millions; ten thousands; thousands; hundreds; tens; units (or ones); tenths; hundredths factor; multiple; prime number; prime factor; composite number; common multiple; >; greater than; >; less than; approximately; estimate; halfway between; number sequence; term

1A/B The number system and place value

Discover and Explore

Write the number given in each question in the place-value grid and say it aloud.

Write the column headings in the place-value grid before you write the number.

Example:
8 596 854 people live in Bangkok

Millions	Hundred thousands	Ten thousands	Thousands	Hundreds	Tens	Units	.	Tenths	Hundredths
8	5	9	6	8	5	4	.	0	0

This number is: *Eight million, five hundred and ninety-six thousand, eight hundred and fifty-four*

1. The circumference of the Earth is 39 775.71 km.

							.		
							.		

2. The world's population grew by 1 665 671 in the first week of May 2014.

							.		
							.		

1A/B The number system and place value

Discover and Explore

3. In a typical month 5 625 429 cars are produced.

							.		

4. There are 6 227 678 computers sold every week.

							.		

5. Every minute 415 151 tweets are sent.

							.		

6. The current world record for the 100 m sprint is 9.58 seconds. It is held by Usain Bolt.

							.		

1C Number properties

Discover

A prime number has only two factors: itself and 1.
17 is a prime number because its only factors are 17 and 1.

A prime factor is a factor which is a prime number.

You can use a factor tree to find the prime factors of a number.
For example:
9 and 4 are not prime numbers but
we can split them into factors which are
prime numbers.
3 and 2 are both prime numbers so the prime factors of
36 are 3 × 3 × 2 × 2.

```
      36
     /  \
    9    4
   / \  / \
  3  3 2  2
```

Draw factor trees to find the prime factors of these numbers.

1. 60   **5.** 36   **9.** 87

2. 28   **6.** 72   **10.** 50

3. 85   **7.** 90

4. 52   **8.** 64

1C Number properties

Explore

Multiples of 3 are 3, 6, 9, 12, 15, 18, 21, 24, 27, ...
Multiples of 12 are 12, 24, 36, 48, 60, 72, ...

12 and 24 appear in both lists, so they are **common multiples** of 3 and 12. The smallest of these common multiples is 12, so 12 is the **lowest common multiple**. Find the lowest common multiple of each pair of numbers by writing the lists of multiples.

1. What is the lowest common multiple of 4 and 15?

2. What is the lowest common multiple of 5 and 11?

3. What is the lowest common multiple of 6 and 16?

4. What is the lowest common multiple of 7 and 13?

5. What is the lowest common multiple of 6 and 19?

6. What is the lowest common multiple of 7 and 14?

7. What is the lowest common multiple of 8 and 12?

8. What is the lowest common multiple of 9 and 17?

1D/E Comparing numbers and estimation and rounding

Discover

Use the digits 3, 5, 6 and 9 to make a 4-digit number. Write your 4-digit number on a number line. Then round it to the nearest thousand. Repeat six times.

Example:
My number is <u>5693</u>.

It is between <u>5000</u> and <u>6000</u> but nearer to <u>6000</u>.
So my number is <u>6000</u> to the nearest thousand.

1. My number is _____.

   It is between _____ and _____ but nearer to _____.
   So my number is _____ to the nearest thousand.

2. My number is _____.

   It is between _____ and _____ but nearer to _____.
   So my number is _____ to the nearest thousand.

3. My number is _____.

   It is between _____ and _____ but nearer to _____.
   So my number is _____ to the nearest thousand.

4. My number is _____.

   It is between _____ and _____ but nearer to _____.
   So my number is _____ to the nearest thousand.

5. My number is _____.

   It is between _____ and _____ but nearer to _____.
   So my number is _____ to the nearest thousand.

1D/E Comparing numbers and estimation and rounding

Explore

Use the digits 5, 6 and 7 for this activity.

1. Write all the different 3-digit numbers you can make using these digits.

2. Write the numbers in order in the table, starting with the smallest. Then complete the table by rounding each number to the nearest 10 and the nearest 100. One is done for you.

   Remember that we **round up** to the nearest 10 if the number ends in 5.

Number	Number rounded to the nearest 10	Number rounded to the nearest 100
567	570	600

3. Fill in the blanks using a number from the table.

   Remember: < means less than
   > means greater than

   a. _____ > 750

   b. _____ < 580

   c. _____ < 650

   d. _____ > 695

   e. _____ < 650 < _____

   f. _____ > 700 > _____

1F Number sequences

Discover

Create your own number sequences. Follow these steps to make the rules for each sequence.

1. Think of a starting number.
2. Think of an operation (+, −, ×, ÷).
3. Think of a number to use each time.

Example:

Starting number	Operation	Number to use	Sequence
15	subtract (−)	7	15 8 1 −6 −13 −20 ...

Write your number sequences in the table. Write something you notice about each sequence.

	Starting number	Operation	Number to use	Sequence
1.				
	I notice that			
2.				
	I notice that			
3.				
	I notice that			
4.				
	I notice that			
5.				
	I notice that			
6.				
	I notice that			
7.				
	I notice that			
8.				
	I notice that			

1F Number sequences

Explore

Look at these consecutive numbers.

3   4   5   6

Write + and − signs between the numbers. Find all the different possible arrangements.

Example: 3 + 4 − 5 − 6 = −4

3   4   5   6 = ☐

3   4   5   6 = ☐

3   4   5   6 = ☐

3   4   5   6 = ☐

3   4   5   6 = ☐

3   4   5   6 = ☐

Repeat the activity with the numbers 5, 6, 7, 8.

5   6   7   8 = ☐

5   6   7   8 = ☐

5   6   7   8 = ☐

5   6   7   8 = ☐

5   6   7   8 = ☐

5   6   7   8 = ☐

1 Review

Write five facts about each of the numbers below. Use each word in the box at least once.

> million   ten thousand   thousand   hundred
> units   tenths   hundredths
> factor   multiple   prime number   prime factor   common multiple
> is greater than   is less than   approximately   halfway between

38

7 654 321

9.52

17.65

2 Fractions and Decimals

Introduction

The most important thing for students to understand about fractions is the idea of 'equality'. They need to understand that fractions are 'equal areas' or 'equal shares'. This means that when we calculate with fractions we have to divide into equal parts. From their work in earlier stages, students are now used to how fractions are written.

$$\frac{2}{9}$$

$2 \leftarrow$ This is the numerator and tells us the number of parts in the fraction.

$9 \leftarrow$ This is the denominator and tells us how many equal parts the whole is divided into.

This unit also develops students' understanding of improper fractions, which are fractions greater than 1. These can be written as either improper fractions or mixed numbers.

For example: $\frac{7}{4}$ is an improper fraction. The equivalent mixed number is $1\frac{3}{4}$.

In this unit students use their knowledge of fractions to calculate ratios and proportions.

Ratio compares part to part. For example: 1 to 4 is a ratio.
There is one adult to every four children on a school trip. The ratio of adults to children is 1:4. The ratio of children to adults is 4:1.

Proportion compares a part to the whole. For example: 1 out of 5 is a proportion.
One out of every five people on the trip is an adult. Four out of five people on the trip are children. Proportions can be expressed as fractions. $\frac{1}{5}$ of the people on the trip are adults. $\frac{4}{5}$ of the people on the trip are children. Proportions are often expressed as percentages. If $\frac{1}{5}$ of the people are adults we need to work out $\frac{1}{5}$ as a percentage. $\frac{1}{5}$ of 100 is 20 so $\frac{1}{5}$ is 20%. (To convert a fraction to a percentage, divide 100 by the denominator and multiply by the numerator.)

Ways to help

As in the previous stages, students still need to see fractions modelled practically. Students can make a fraction wall like this one to display on the wall at home. They could write the equivalent decimal fractions and percentages on the appropriate sections of the wall.

Key Words

part; whole; fraction; equivalent fraction; mixed number; improper fraction; numerator; denominator; half; third; quarter; fifth; sixth; eighth; tenth; equivalent fraction; simplest form; decimal number; decimal fraction; percentage; ratio; proportion

2A Equivalent fractions

Discover

Use this fraction wall to help you find equivalent fractions.

For each fraction, write as many different equivalent fractions as you can.

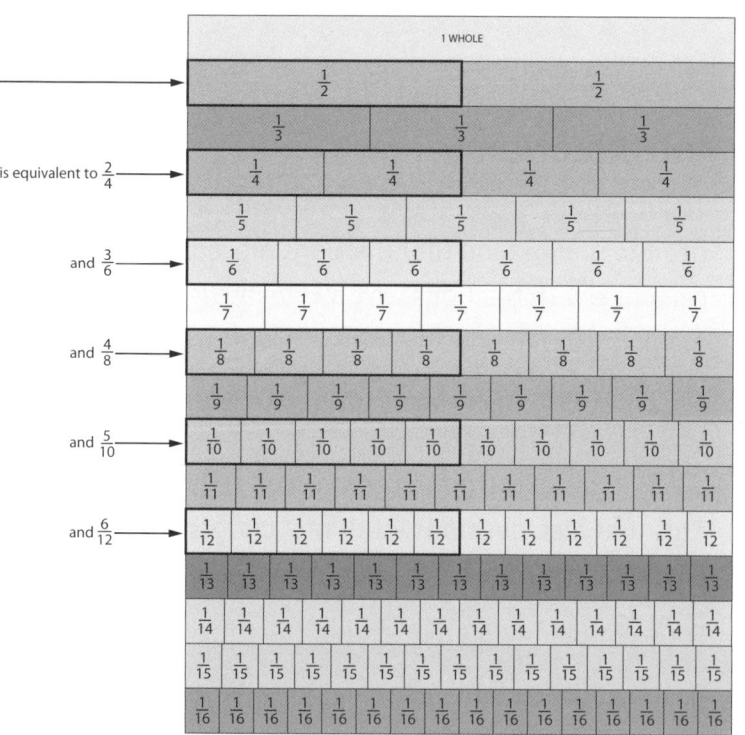

$\frac{1}{2}$ is equivalent to $\frac{2}{4}$ and $\frac{3}{6}$ and $\frac{4}{8}$ and $\frac{5}{10}$ and $\frac{6}{12}$

$\frac{2}{3}$	$\frac{3}{5}$	$\frac{5}{6}$	$\frac{1}{8}$
$\frac{4}{10}$	$\frac{2}{12}$	$\frac{10}{15}$	$\frac{12}{15}$

2A Equivalent fractions

Explore

Use this fraction wall to help you order fractions.

Cancel the fractions down to their simplest form before writing the answer.

Example: $\dfrac{3}{6}$ and $\dfrac{2}{12}$

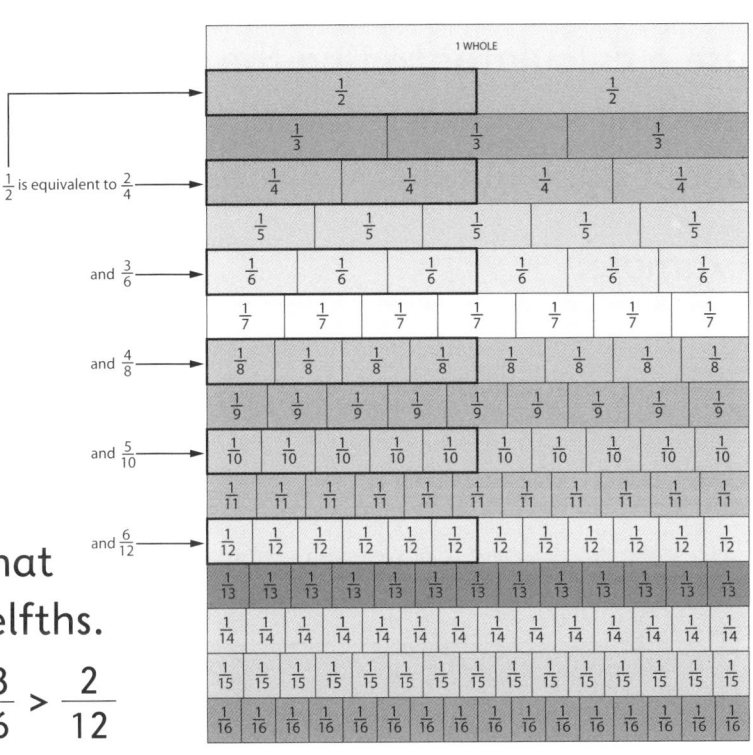

To order these fractions, first change the denominators so that both are sixths or both are twelfths.

$\dfrac{3}{6} = \dfrac{6}{12}$ or $\dfrac{2}{12} = \dfrac{1}{6}$ So $\dfrac{3}{6} > \dfrac{2}{12}$

Then cancel the fractions down.

$\dfrac{1}{2} > \dfrac{1}{6}$

1. $\dfrac{3}{4}$ and $\dfrac{3}{8}$

2. $\dfrac{2}{5}$ and $\dfrac{5}{10}$

3. $\dfrac{8}{14}$ and $\dfrac{3}{7}$

4. $\dfrac{9}{13}$ and $\dfrac{4}{9}$

5. $\dfrac{5}{15}$ and $\dfrac{3}{6}$

6. $\dfrac{8}{16}$ and $\dfrac{3}{8}$

2B Fractions and decimals

Discover

Use a calculator to find the decimal equivalent of each fraction by dividing the numerator by the denominator. Write what you notice about each family.

Example:

Fraction	$\frac{1}{3}$	$\frac{2}{3}$	$\frac{3}{3}$
Decimal equivalent	0.3333333333	0.66666666666	1

The decimals are recurring. 2/3 is double 1/3.

1.

Fraction	$\frac{1}{7}$	$\frac{2}{7}$	$\frac{3}{7}$	$\frac{4}{7}$	$\frac{5}{7}$	$\frac{6}{7}$	$\frac{7}{7}$
Decimal equivalent							

2.

Fraction	$\frac{1}{9}$	$\frac{2}{9}$	$\frac{3}{9}$	$\frac{4}{9}$	$\frac{5}{9}$	$\frac{6}{9}$	$\frac{7}{9}$	$\frac{8}{9}$	$\frac{9}{9}$
Decimal equivalent									

3.

Fraction	$\frac{1}{12}$	$\frac{2}{12}$	$\frac{3}{12}$	$\frac{4}{12}$	$\frac{5}{12}$	$\frac{6}{12}$	$\frac{7}{12}$	$\frac{8}{12}$	$\frac{9}{12}$	$\frac{10}{12}$	$\frac{11}{12}$	$\frac{12}{12}$
Decimal equivalent												

2B Fractions and decimals

Explore

A You will need a set of 0–9 digit cards for this activity. Pick three cards. Arrange the digits to make six different numbers, each with two decimal places. Arrange the numbers in order of size, smallest first.

Example: [2] [7] [1]

My numbers, in order, are: 1.27   1.72   2.17   2.71   7.12   7.21

Repeat the activity five times.

1. My numbers, in order, are: _____

2. My numbers, in order, are: _____

3. My numbers, in order, are: _____

4. My numbers, in order, are: _____

5. My numbers, in order, are: _____

B Continue the sequences in the table by counting on or back.

Instruction	Start number					
Count on in steps of 0.1	0.4	0.5				
Count on in steps of 0.2	0.1	0.3				
Count back in steps of $\frac{1}{4}$	1	$\frac{3}{4}$				$-\frac{1}{4}$
Count back in steps of $\frac{1}{2}$	$1\frac{1}{2}$					
Count on in steps of 0.5	2	2.5				

2C Addition pairs

Discover

Complete the table by using the digits in each number to make the addition pairs to 100, to 10 and to 1. The first one is done for you.

Number	Addition pair to 100	Addition pair to 10	Addition pair to 1
38	38 + 62 = 100	3.8 + 6.2 = 10	0.38 + 0.62 = 1
85			
41			
16			
73			
24			
11			
89			
54			
67			
14			
51			
40			
5			
92			

2C Addition pairs

Explore

1. Write ten addition pairs to 100.

2. Write ten addition pairs to 1.

3. Write ten addition pairs to 10.

4. Write ten addition pairs to 50.

5. Write ten addition pairs to 5.

6. Write ten addition pairs to 20.

7. Write ten addition pairs to 2.

8. Write ten addition pairs to 1000.

2D Mixed numbers and improper fractions

Discover

1. Write the improper fractions as mixed numbers.

   Example: $\frac{13}{3} = 4\frac{1}{3}$ (because $13 \div 3$ is 4 remainder 1)

   a. $\frac{9}{2}$

   b. $\frac{17}{4}$

   c. $\frac{16}{5}$

   d. $\frac{23}{5}$

   e. $\frac{9}{4}$

   f. $\frac{11}{6}$

   g. $\frac{25}{8}$

   h. $\frac{31}{10}$

   Remember: You divide the numerator by the denominator and write the remainder as a fraction.

2. Write the mixed numbers as improper fractions.

   Example: $4\frac{3}{5} = \frac{23}{5}$ (because $4 \times 5 + 3 = 23$)

   a. $3\frac{1}{4}$

   b. $4\frac{1}{8}$

   c. $4\frac{3}{8}$

   d. $5\frac{2}{9}$

   e. $1\frac{9}{10}$

   f. $10\frac{1}{8}$

   g. $6\frac{4}{5}$

   h. $3\frac{9}{11}$

   Remember: You multiply the whole number by the denominator and add the fractional part.

2D Mixed numbers and improper fractions

Explore

Complete the table. Choose a suitable improper faction and equivalent mixed number each time.

The first one is done for you.

	Mixed number	Improper fraction
A fraction bigger than 3	$4\frac{2}{5}$	$\frac{22}{5}$
A fraction bigger than 5		
A fraction smaller than 7		
A fraction between 7 and 9		
A fraction bigger than $5\frac{2}{5}$		
A fraction smaller than $4\frac{3}{4}$		
A fraction bigger than $10\frac{3}{5}$		
A fraction between 8 and 9		
A fraction bigger than 15		
A fraction smaller than 10		
A fraction between $6\frac{1}{2}$ and $6\frac{3}{4}$		

2E Ratio and proportion

Discover and Explore

> Remember that ratio compares part to part.

There are 15 girls in a class of 25. The ratio of girls to boys is 15:10. In its simplest terms we can write this as 3:2. (Divide 15 and 10 by 5 to simplify the ratio.)

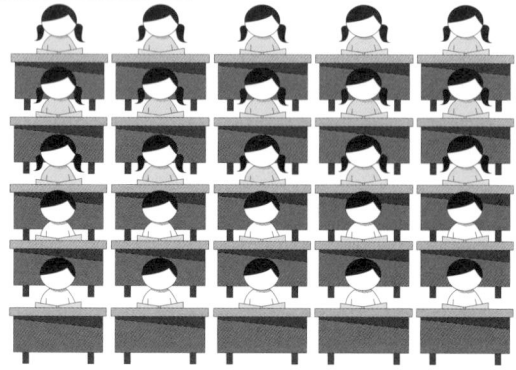

1. What is the ratio of girls to boys in each class? Draw an array to match your answer.

   a. 4 girls in a class of 12

   b. 6 girls in a class of 24

   c. 3 girls in a class of 15

   d. 2 girls in a class of 16

   e. 15 girls in a class of 20

   f. 12 girls in a class of 15

Discover and explore

Remember that proportion compares a part to the whole.

$\frac{3}{10}$ of the grid is shaded. The proportion of the grid that is shaded is 0.3 or 30%.

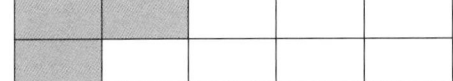

2. Write the proportion of each grid that is shaded. Write the proportion as a fraction, a decimal and a percentage.

   a. Fraction: _____ Decimal: _____ Percentage: _____

   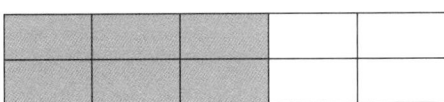

   b. Fraction: _____ Decimal: _____ Percentage: _____

   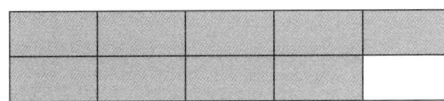

   c. Fraction: _____ Decimal: _____ Percentage: _____

   d. Fraction: _____ Decimal: _____ Percentage: _____

   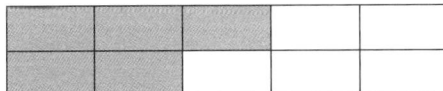

   e. Fraction: _____ Decimal: _____ Percentage: _____

   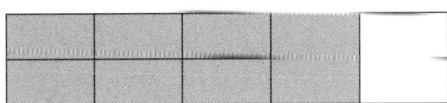

3. Colour each grid to show the proportion of coloured tiles.
   a. 0.75 or 75% coloured tiles

   b. 0.125 or 12.5% coloured tiles

2F Percentages

Discover

Use your knowledge of percentages and their equivalent fractions to answer the questions.

Remember that percentage means 'out of 100'.

1. Write these percentages as fractions.

   a. 25% = ☐   f. 90% = ☐

   b. 75% = ☐   g. 50% = ☐

   c. 10% = ☐   h. 33.33% = ☐

   d. 40% = ☐   i. 66.66% = ☐

   e. 20% = ☐   j. 70% = ☐

2. Complete the table to show the equivalent percentages for marks in a test. The test was marked out of 60.

Mark (out of 60)	Equivalent %
60	100%
54	
45	
42	
30	50%
24	
15	
12	
6	10%
3	

2F Percentages

Explore

Complete the spider diagrams by writing six different percentages of the number in the centre.

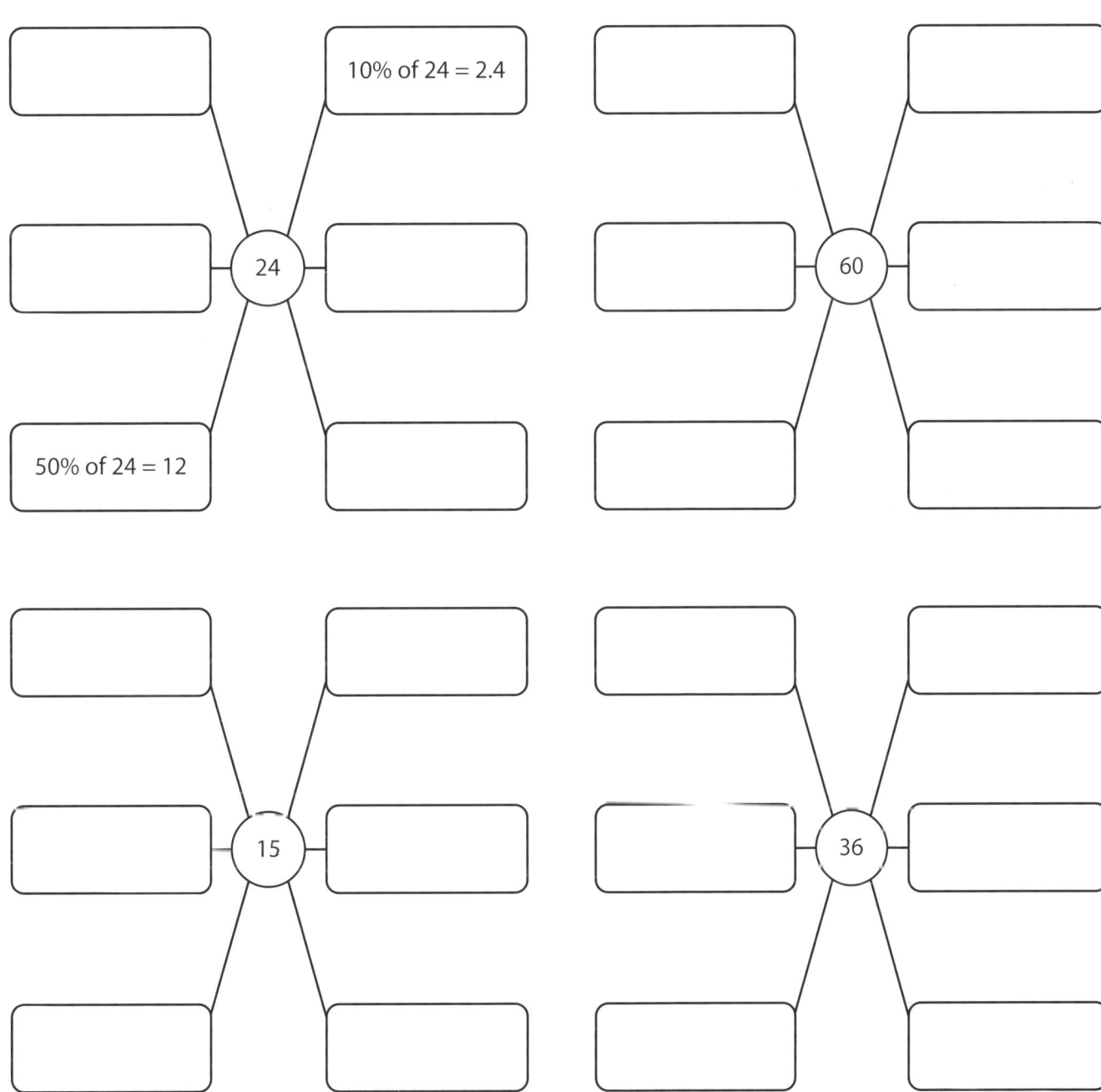

2 Review

Use your knowledge of fractions and percentages to answer the word problems.

1. The shopkeeper reduces the price of all his stock by 50%. What are the new prices?

   T-shirt: ☐   Jeans: ☐

   Trainers: ☐   Socks: ☐

2. I buy 5 pairs of socks. There is a 10% discount for buying 5 pairs.

   a. What is the discount? ☐   b. How much do I pay? ☐

3. My friend says that a T-shirt costs $\frac{1}{3}$ the amount of a pair of jeans. Is she correct? Why?

   ☐

4. I buy one of each item. The shopkeeper gives me $\frac{1}{3}$ off the total price.

   a. What is the original total price? ☐

   b. What is the discount? ☐

   c. How much do I pay? ☐

5. I buy a T-shirt and a pair of jeans. The shopkeeper gives me 10% discount.

   a. What is the original total price? ☐

   b. What is the discount? ☐

   c. How much do I pay? ☐

3 Mental Calculation

Introduction

Mental calculation is one of the most important skills that students can learn. The first response to any problem should be 'Can I calculate that mentally?' Calculators or a pen and paper should be a last resort, and when using these, students should always make an estimate of the answer so that they can check their results.

This unit focuses specifically on how students can use facts that they know to derive new ones. This is at the heart of problem solving. Techniques that they will use include using multiples and near multiples and doubles and near doubles.

Ways to help

As with the units on mental calculation in the earlier stages, the best way to help students is to ask them to explain their strategies to you. Talking about their methods will help them come to a much clearer understanding of how they are approaching calculations, which will allow them to apply the same methods to solve other similar problems. Try to carry out the calculations yourself and discuss with students the strategies that you used. Compare strategies and discuss which is the most efficient in different situations.

Key Words

single-digit number; 2-digit number; 3-digit number; 4-digit number; 5-digit number; sum; product; divisibility; multiple; factor; near multiple; double; near double; halve; array; row; column; divisor; remainder

3A/F Mental strategies for addition and subtraction

Discover

Complete the calculations.
Work with a partner. Tell each other your strategy. Write your strategies down. If you use the same strategy write 'same strategy'.

I know that 64 + 20 is 84. If I subtract 1, I know that 64 + 19 is 83.

Calculation	My strategy	My partner's strategy
64 + 19		
5.6 + 3.9		
209 – 199		
998 + 15		
2999 + 3156		
25.8 – 13.2		
5425 – 1998		
295 + 66		
1999 + 2999		
29.6 – 5.1		
507 – 10		
77 + 7		

3A/F Mental strategies for addition and subtraction

Explore

For each answer write five different questions.
They should involve 2- and 3-digit numbers. Include a mixture of addition and subtraction questions.
Give your questions to a friend to solve.

1. The answer is 250.

   a. _____

   b. _____

   c. _____

   d. _____

   e. _____

2. The answer is 1001.

   a. _____

   b. _____

   c. _____

   d. _____

   e. _____

3. The answer is 98.

   a. _____

   b. _____

   c. _____

   d. _____

   e. _____

4. The answer is 999.

   a. _____

   b. _____

   c. _____

   d. _____

   e. _____

3B/E Mental strategies for multiplication and division

Discover

You will need a set of 1–9 digit cards for this activity.

To complete each row of the table, pick five digit cards. Use these five digits to make a number. Use your number to make a division calculation, following the instruction in each row of the table.
You do not have to use all the cards, but challenge yourself to use as large a number as possible.

	Calculation	Answer
Division by 2 with a remainder		
Division by 2 with no remainder		
Division by 3 with a remainder		
Division by 3 with no remainder		
Division by 4 with a remainder		
Division by 4 with no remainder		
Division by 5 with a remainder		
Division by 5 with no remainder		
Division by 6 with a remainder		
Division by 6 with no remainder		
Division by 7 with a remainder		
Division by 7 with no remainder		
Division by 8 with a remainder		
Division by 8 with no remainder		
Division by 9 with a remainder		
Division by 9 with no remainder		
Division by 10 with a remainder		
Division by 10 with no remainder		

3B/E Mental strategies for multiplication and division

Explore

Complete these multiplication grids.

1.

×	20	40	30	60	90
5					
60					
70					
10					
50					

2. You can use your answers to question 1 to help you with this one.

×	20	40	30	60	90
9					
61					
69					
11					
49					

3.

×	16	25	50	30	60
0.4					
0.8					
0.2					
0.5					
0.3					

3C Using known facts to derive new ones

Discover

You can work out multiplication facts for 13 and for 19 using facts that you already know. Complete the tables. Two have been done for you.

Multiplication	Answer	Reason
1 × 13		
2 × 13		
3 × 13		
4 × 13		
5 × 13		
6 × 13		
7 × 13	91	because 7 × 10 = 70 and 7 × 3 = 21 70 + 21 = 91
8 × 13		
9 × 13		
10 × 13		

Multiplication	Answer	Reason
1 × 19		
2 × 19		
3 × 19		
4 × 19		
5 × 19	95	because 5 × 10 = 50 and 5 × 9 = 45 50 + 45 = 95
6 × 19		
7 × 19		
8 × 19		
9 × 19		
10 × 19		

3C Using known facts to derive new ones

Explore

If you know that 80 × 7 = 560, you also know that 80 × 0.7 = 56 and 40 × 7 = 280.

Complete the two spider diagrams using the facts in the centre to help you.

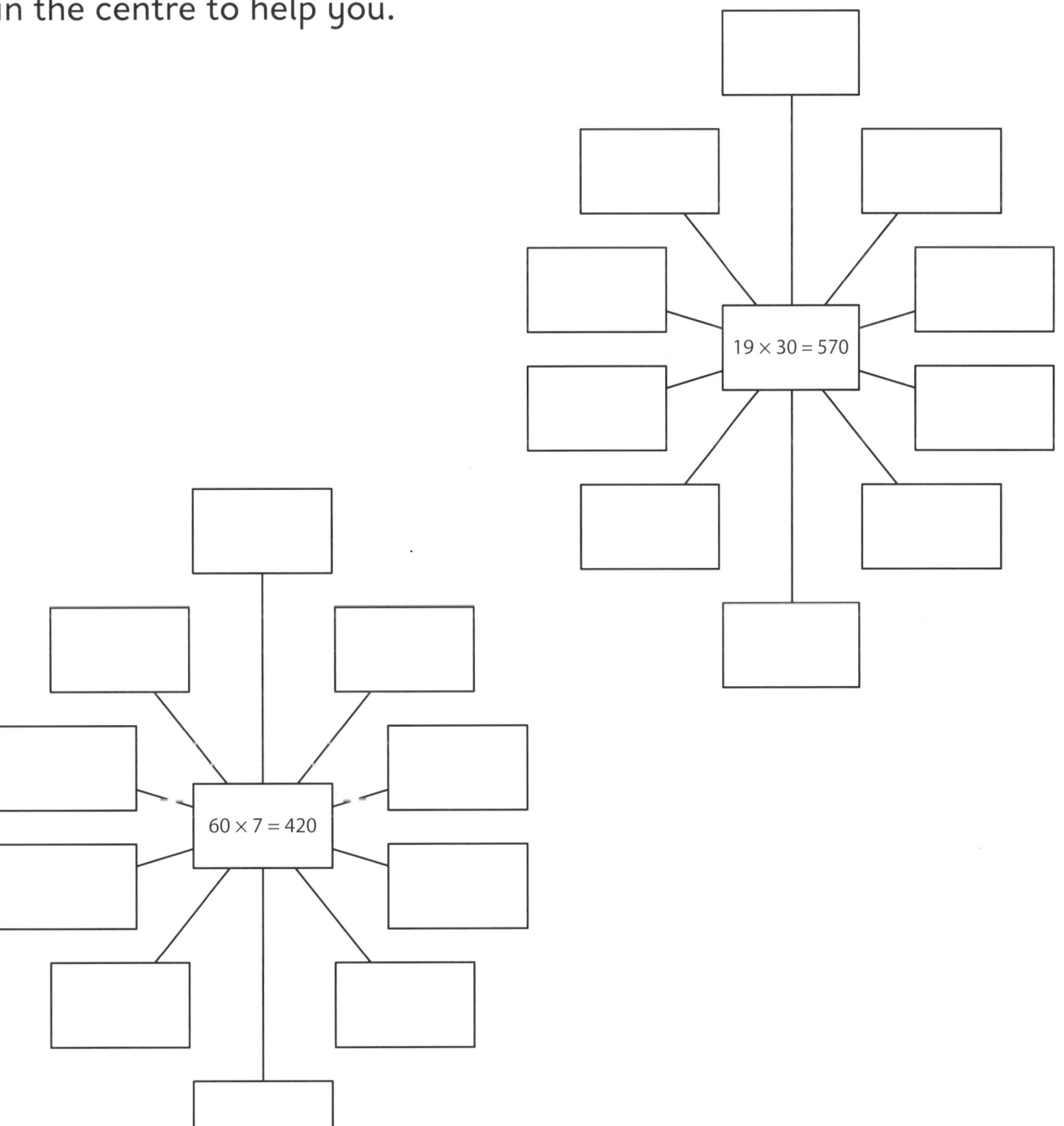

3D Doubling and halving

Discover

You will need a set of 1–9 digit cards for this activity.

Pick two cards and make two different 2-digit numbers. Use these numbers as your starting numbers to complete the table.

Repeat the activity by picking two more cards from the cards you have left. You will be able to do the activity four times altogether, until there is just one card left.

Example: *I picked 7 and 2.*

Starting number	Double	Divide by 10 and double	Divide by 100 and double
72	Double 72 = 144	Double 7.2 = 14.4	Double 0.72 = 1.44
27	Double 27 = 54	Double 2.7 = 5.4	Double 0.27 = 0.54

Complete this table.

Starting number	Double	Divide by 10 and double	Divide by 100 and double

3D Doubling and halving

Explore

This is a price list at the supermarket.

Calculate the cost of each shopping list.
Use the squared paper for your working out.

Items	Price
4 bananas	$1.50
2 mangos	$0.80
6 tomatoes	$2.10
2 packs grapes	$1.24

1.
- 8 bananas
- 1 mango
- 3 tomatoes
- 1 pack grapes

Total cost _____

2.
- 2 bananas
- 2 mangos
- 6 tomatoes
- 4 packs grapes

Total cost _____

3.
- 4 bananas
- 4 mangos
- 3 tomatoes
- 4 packs grapes

Total cost _____

3 Review

Use these digits to make up a set of calculations.

| 1 | 3 | 5 | 6 | 9 | 0 |

You do not have to use all the digits for each question but you can use each digit more than once. There are lots of possible answers for each question.

Example: A question with the answer 20, involving numbers with one place of decimals — 9.9 + 10.1

1. Another question with the answer 20, involving numbers with one place of decimals

2. A question with the answer 10, involving numbers with one place of decimals

3. An addition question involving a near multiple of 100

4. A subtraction question involving a near multiple of 100

5. An addition question involving two 2-digit numbers

6. A subtraction question involving two 2-digit numbers

7. A multiplication question involving a near multiple of 10

8. A question involving doubling a number with two places of decimals

4 Addition and Subtraction

Introduction

The last unit focused on mental calculate strategies, whereas this unit teaches students formal written methods that they can use to carry out calculations which cannot be done mentally. Learning the different calculation methods allows students to decide when they can calculate something mentally and when they need to use a written method (or algorithm). Even when they use a formal written method, they should always estimate the answer first so that they can check the accuracy of their working out.

At Stage 6, students will continue to use partitioning to carry out calculations and will use column methods to calculate with 3-digit numbers and with money. They will also calculate with negative numbers in the context of temperature.

Ways to help

Some of the methods that students use may not be the methods that you were taught at school. Do not worry about this. Asking students to explain to you how the methods work will help them understand the methods better. It may also mean that you discover new and more effective ways of adding and subtracting.

> **Key Words**
>
> partition; tens boundary; hundreds boundary; positive; negative; directed numbers; above zero; below zero; units; ones; tens; hundreds; thousands; tenths; hundredths; decimal point

4A Adding and subtracting 3-digit numbers

Discover

In this activity you will practise adding and subtracting 3-digit numbers.

Solve the following calculations. Decide if you can work each one out mentally or if you need to use a written method.
- If you use a written method, show all your working out.
- If you calculate the answer mentally, write your strategy.

175.6 + 28.3	398 + 46
119.7 − 18.4	965.2 + 425.8
210 + 698	478.6 − 139.8
156 + 999	98.6 − 44.5
99.6 − 89.9	56.88 + 98.75

4A Adding and subtracting 3-digit numbers

Explore

You will need a set of 0–9 digit cards for this activity.
For each challenge pick six cards.
Use your digits to write an addition and a subtraction with answers that fit the challenge.
Choose whether to use four, five or six of the digits, but you must use at least four.
You can use the same digit twice.

Example:

My digits	Challenge	Addition	Subtraction
1, 7, 4, 8, 9, 0	As close to 10 as you can	9.1 + 0.9 = 10	14.8 − 4.8 = 10

Complete this table.

My digits	Challenge	Addition	Subtraction
	As close to 10 as you can		
	As close to 100 as you can		
	As close to 1000 as you can		
	As close to 750 as you can		
	As close to 1 as you can		
	As close to 75 as you can		
	As close to 15.8 as you can		
	As close to 250 as you can		
	As close to 19.8 as you can		

4B Adding and subtracting money

Discover and Explore

In these word problems you will solve calculations involving money. Read each problem carefully to decide which operation you should use.

Show all of your working for each question.

1. I visit a museum with my family and a friend. The family ticket is $28.55. My friend's ticket costs $19.79. What is the total cost?

   _____

2. My friend and I want to buy a game. The game costs $25. I have $14.67 and my friend has $8.88. Can we afford the game if we put our money together?

   _____

3. I need to buy some new clothes for my holiday. I buy two T-shirts, which cost $8.15 each. I also buy a pair of shorts, which cost $9.64, and a cap, which costs $2.87. How much do I spend altogether?

   _____

4B Adding and subtracting money

Discover and Explore

4. I go shopping with $25. How much change will I get if I buy the items in question 3?

5. You are planning a class trip. You have collected $175 from the students. The transport costs $88.50 and the entrance to the attraction costs $65.95. How much do you have left to buy refreshments?

6. Three classes in your school are taking part in a sponsored swim. The three classes raise $42.78, $39.83 and $56.13. How much do they raise altogether? Your headteacher has promised to make the amount up to $150. How much does the headteacher have to pay?

4C Using negative numbers

Discover and Explore

This table shows the temperatures, in degrees Celsius, during a year in Bangkok, Moscow, Reykjavik and the North Pole.

	Jan	Feb	Mar	Apr	May	Jun	Jul	Aug	Sep	Oct	Nov	Dec
Bangkok	21	23	25	27	25	23	25	23	23	23	21	20
Moscow	−7	−8	−1	7	15	18	21	19	13	6	0	−4
Reykjavik	−5	−3	−3	2	7	8	9	7	5	4	0	−3
North Pole	−25	−19	−14	−10	0	9	13	11	0	−17	−23	−28

1. Draw the warmest and coldest temperatures for each location on these pairs of thermometers.

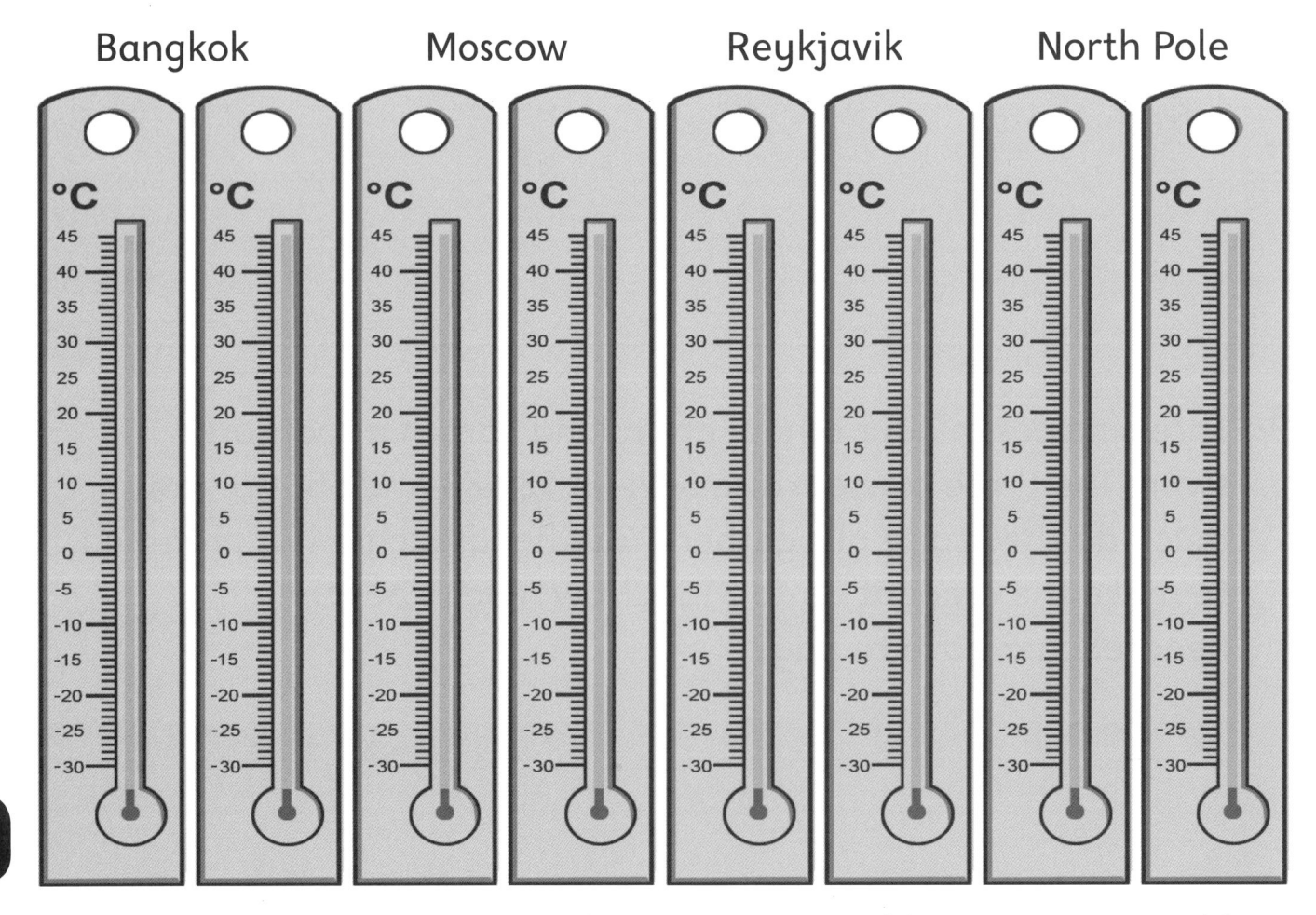

Bangkok   Moscow   Reykjavik   North Pole

Warmest Coldest Warmest Coldest Warmest Coldest Warmest Coldest

4C Using negative numbers

Discover and Explore

2. Use the temperature table to find the differences given below. For each question draw a number line to show how you calculated the answer.

a. The difference between the coldest and warmest month in Bangkok.
☐ degrees Celsius

b. The difference between the coldest and warmest month in Moscow.
☐ degrees Celsius

c. The difference between the coldest and warmest month in Reykjavik.
☐ degrees Celsius

d. The difference between the coldest and warmest month at the North Pole.
☐ degrees Celsius

e. The difference between the coldest month in Moscow and the warmest month in Bangkok.
☐ degrees Celsius

f. The difference between the coldest month at the North Pole and the warmest month in Moscow.
☐ degrees Celsius

g. The difference between the coldest month in Bangkok and the warmest month in Reykjavik.
☐ degrees Celsius

h. The difference between the coldest month in Moscow and the coldest month in Bangkok.
☐ degrees Celsius

4 Review

Use all the strategies that you have learned in this unit to answer the questions.

1. I am 1.78 m tall. My friend is 26 cm taller than me. How tall is my friend?

2. Another friend is 29 cm shorter than me. How tall is this friend?

3. What is the total height of me and my two friends?

4. A tree in the local park is 7.5 m tall. How much taller is the tree than the total of our heights?

5. I buy a notebook, two pencils and a pencil case. The total amount I spend is $10.55. What prices could each of the three items be? What are two other possible prices for each of the items?

6. Which two numbers are 15 steps from −7 on the number line?

7. Which two numbers are 11 steps from +6 on the number line?

8. At dawn the temperature is −11. It is 8 degrees warmer at midday. What is the temperature at midday?

5 Multiplication and Division

Introduction

This unit develops students' mental and informal methods of multiplication and division. They then apply this knowledge when using formal methods to calculate using pencil and paper. In this unit they develop their understanding to calculate with 2-, 3- and 4-digit numbers. Students will carry out division calculations that involve remainders and learn the arithmetical laws that apply to multiplication and division.

Ways to help

Students should always estimate before starting to calculate. When you are helping students, before they start to answer a question always ask what they estimate the answer to be. You should also ask them to explain how the methods that they are using work. This will help students gain a better understanding of the methods (or algorithms), allowing them to use them to solve a wide range of problems.

Key Words

multiple; product; factor; partition; estimate; double; halve; share equally; divide; divisor; dividend; quotient; remainder

5A Multiplying by 2-, 3- and 4-digit numbers

Discover and Explore

In these questions you will practise the different strategies for solving multiplication calculations.

1. Answer these questions by partitioning. The first one is done for you.

   a. 486 × 4 = 400 × 4 + 80 × 4 + 6 × 4 = 1600 + 320 + 24 = 1944

   b. 532 × 6 _____

   c. 815 × 3 _____

   d. 268 × 9 _____

   e. 381 × 5 _____

   f. 659 × 7 _____

2. Use the grid method to solve these multiplications. The first one is done for you.

   a. 384 × 26

	20	6	
300	6000	1800	7800
80	1600	480	2080
4	80	24	104
			9984

   Answer: 384 × 26 = 9984

   b. 262 × 53

   Answer: _____

   c. 648 × 36

   Answer: _____

   d. 439 × 24

   Answer: _____

5A Multiplying by 2-, 3- and 4-digit numbers

Discover and Explore

3. Use the column method of multiplication to solve the calculations.

Example: 258 × 53

Estimate first: The answer will be approximately 250 × 50 = 12500

```
    258
  ×  53
  -----
  12900    (258 × 50 = 12900)
    774    (258 × 3 = 774)
  -----
  13674
```

a. 298 × 46

b. 315 × 27

c. 586 × 39

d. 814 × 19

5B Dividing 3-digit numbers by 2-digit numbers

Discover

Use the written method to solve the division calculations. Make sure that you show all your working out.

Example: 912 ÷ 16

Write the dividend inside the box and the divisor outside.

```
16 ) 912
     800    16 × 50  (First estimate 16 × 50 = 800)
     ---
     112             (Subtract 912 − 800 = 112)
      80    16 × 5   (I know that 16 × 5 = 80)
     ---
      32    16 × 2   (Subtract 112 − 80 = 32, then I know that 2 × 16 = 32)
```

912 ÷ 16 = 50 + 5 + 2 = 57

1. 544 ÷ 17

2. 882 ÷ 14

3. 754 ÷ 13

4. 684 ÷ 18

5B Dividing 3-digit numbers by 2-digit numbers

Explore

Solve these problems. Make sure that you show all your working out. You should estimate first.

1. The answer to a division question is 39. Use the digits 1, 1, 2, 8 and 9 to find the 3-digit dividend and the 2-digit divisor. Show you are correct using a written method.

2. The answer to a division question is 52. Use the digits 1, 4, 7, 8 and 8 to find the 3-digit dividend and the 2-digit divisor. Show you are correct using a written method.

3. The answer to a division question is 63. Use the digits 1, 4, 5, 5 and 9 to find the 3-digit dividend and the 2-digit divisor. Show you are correct using a written method.

4. The answer to a division question is 71. Use the digits 1, 2, 3, 3 and 9 to find the 3-digit dividend and the 2-digit divisor. Show you are correct using a written method.

5C Division with remainders

Discover

For each of these calculations, give your answer as a mixed number and as a decimal.

Remember:
$\frac{1}{10} = 0.1$ $\frac{1}{5} = 0.2$ $\frac{1}{3} = 0.333$
$\frac{1}{4} = 0.25$ $\frac{1}{8} = 0.125$

Example: $56 \div 5 = 11$ remainder 1
$= 11\frac{1}{5} = 11.2$

1. $76 \div 10 =$

2. $157 \div 2 =$

3. $43 \div 3 =$

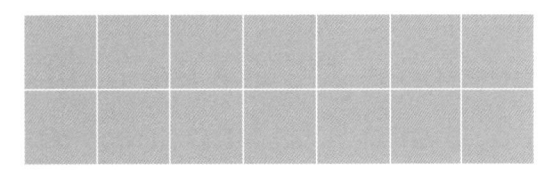

4. $83 \div 8 =$

5. $69 \div 6 =$

6. $128 \div 5 =$

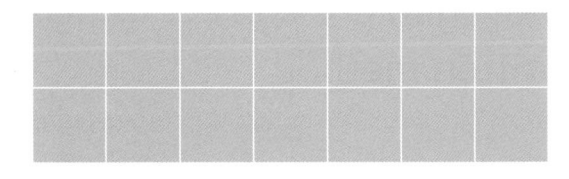

7. $56 \div 10 =$

8. $327 \div 5 =$

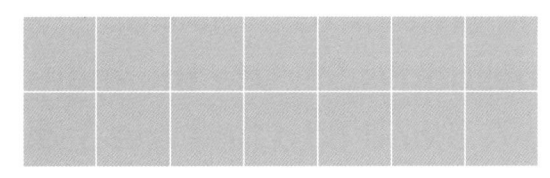

9. $99 \div 8 =$

10. $117 \div 4 =$

5C Division with remainders

Explore

These questions will all leave a remainder. You will need to round each answer as instructed below.

Remember: If the answer finishes in 0.5 or 0.05 you round up. So $5.65 to the nearest 10 cents is $5.70.

1. A pack of 10 cartons of juice costs $8.94. How much does each carton of juice cost to the nearest 10 cents?

2. A pack of 4 cartons of juice costs $3.78. How much does each carton of juice cost to the nearest 10 cents?

3. Approximately how much do I save per carton of juice if I buy a pack of 10 rather than a pack of 4?

4. Washing powder costs $5.85 for 5 kg. How much does 1 kg cost exactly?

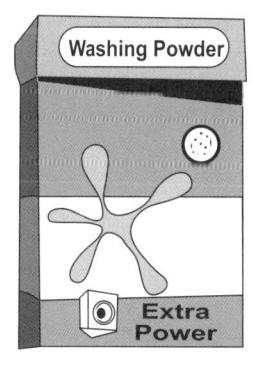

5. I can buy a 2 kg box of washing powder for $2.53. How much does 1 kg cost? Give your answer to the nearest 10 cents.

5D Using the arithmetical laws for multiplication and division

Discover

Change the order of the numbers in these calculations to make them easier to solve. Then work out the answers.

Example: 20 × 18 × 5

　　　　is the same as: 20 × 5 × 18

　　　　= 100 × 18 = 1800

1. 17 × 25 × 4

2. 50 × 26 × 2

3. 30 × 14 × 3

4. 5 × 36 × 5

5. 20 × 16 × 5

6. 5 × 38 × 2

7. 14 × 60 × 10

8. 19 × 25 × 4

9. 8 × 15 × 2

10. 4 × 29 × 25

11. 2 × 38 × 50

12. 12 × 14 × 5

5D Using the arithmetical laws for multiplication and division

Explore

Use the distributive law to carry out these calculations.

Example: 62 × 9

is the same as: 60 × 9 + 2 × 9

= 540 + 18 = 558

1. 74 × 3

2. 86 × 5

3. 29 × 6

4. 36 × 7

5. 82 × 4

6. 99 × 7

7. 84 × 3

8. 26 × 8

9. 37 × 8

10. 46 × 7

11. 37 × 8

12. 82 × 6

5 Review

Your class is planning a trip to the Aquarium. There are 26 students going on the trip. These are the costs.

Train fare: $4.25 per person
Entrance to the Aquarium: $8.40 per person
Refreshments: $338 for 26 people

1. What is the total amount your class will spend on the train fare?

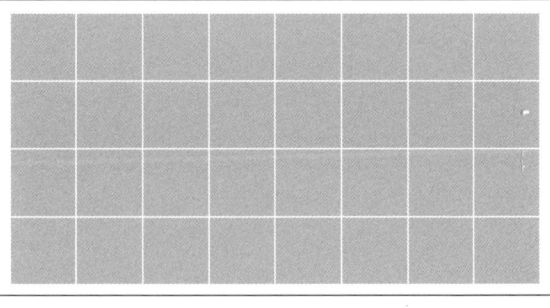

2. What is the total amount your class will spend on entrance to the Aquarium?

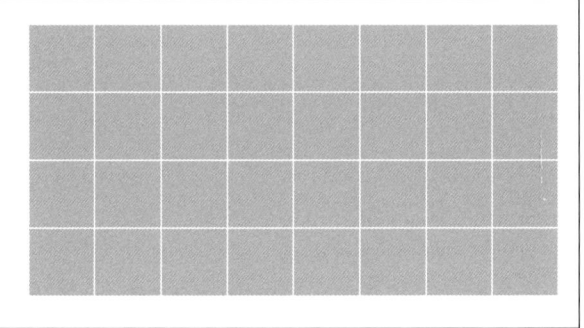

3. How much will your class spend per person on refreshments?

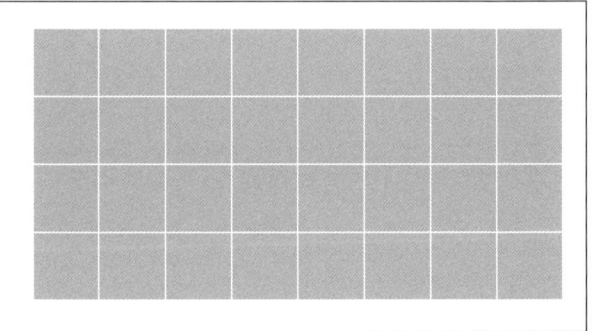

4. How much in total will your class spend per person?

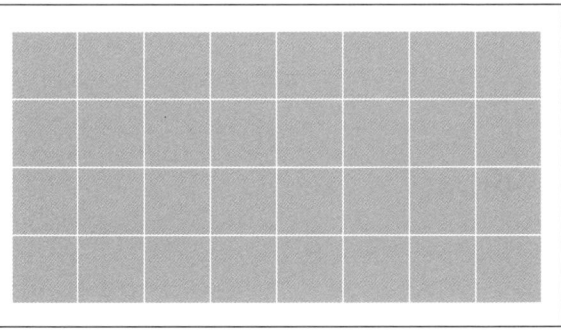

6 Shapes and Geometry

Introduction

In this unit students build on their work in earlier stages on classifying 2-dimensional (2D) and 3-dimensional (3D) shapes using their properties. They will make 2D and 3D shapes and then compare them. By this stage students should be able to name all the common 2D and 3D shapes.

They will also learn how to recognise and draw different angles, including acute, obtuse and reflex angles.

Ways to help

As in the units on shape in earlier stages, the best way to help students is to notice, name and talk about all the shapes that you see in the environment around you. There will be lots of different shapes around the home and in the local area. When you go on visits to new places, take photographs of the shapes you see so that you can talk about their properties later.

Also encourage students to notice symmetry in objects around them, for example in nature, architecture, art and objects in the home.

Key Words

polyhedron; cube; cuboid; pyramid; sphere; hemisphere; cone; cylinder; prism; tetrahedron; triangular prism; octahedron; dodecahedron; circle; semicircle; ellipse; triangle; equilateral triangle; isosceles triangle; scalene triangle; square; rectangle; parallelogram; rhombus; oblong; pentagon; hexagon; heptagon; octagon; polygon; quadrilateral; edges; vertices; faces; curved surfaces; parallel; perpendicular; regular; irregular; diagonal; horizontal; vertical; reflective symmetry; front elevation; side elevation; plan view; isometric drawing; acute angle; obtuse angle; reflex angle

6A Classifying polygons

Discover and Explore

You will need some paper and a pair of scissors for this activity.

Take one piece of paper. Fold the paper so that it is flat. You can make folds at different angles if you like. You can fold the paper as many times as you like.
Make **one** straight cut through the paper so that your paper is separated into different pieces.

Unfold the pieces. How many different polygons do you have? Can you join these together to form different polygons?

> Remember that a polygon is a 2D shape with at least three straight sides.

In the table, sketch five of the different polygons that you have made. Name each polygon and write two of its properties.

Sketch of polygon	Name of polygon	Property 1	Property 2

6B Properties of 3D shapes

Discover and explore

Find six different examples of polyhedra in the environment. Include prisms and pyramids.

Remember that a polyhedron (plural: polyhedra) is a 3D shape with flat faces.

Sketch or stick in a photograph of each shape. Name the shape and write the number of faces, vertices and edges that it has.

Name of shape with sketch or photograph	Number of ...			Name of shape with sketch or photograph	Number of ...		
	faces	vertices	edges		faces	vertices	edges

6C/D Making 2D representations of 3D shapes and angles in a triangle

Discover

You will need a selection of boxes from your home for this activity.

Make a new 3D shape by joining together two different boxes. Draw your new shape. Then draw the plan, the front elevation and the side elevation.

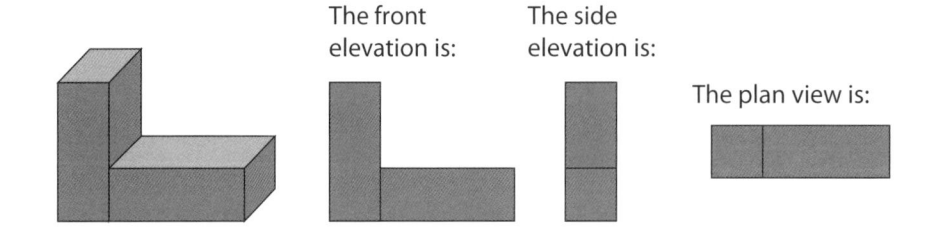

The front elevation is:

The side elevation is:

The plan view is:

Repeat, choosing two different boxes to make another new 3D shape.

Model 2 | Plan
Front elevation
Side elevation

Model 3 | Plan
Front elevation
Side elevation

6C/D Making 2D representations of 3D shapes and angles in a triangle

Explore

Use the pin boards to draw 12 different triangles. Include some isosceles triangles, some scalene triangles and some right-angled triangles. Here are some examples.

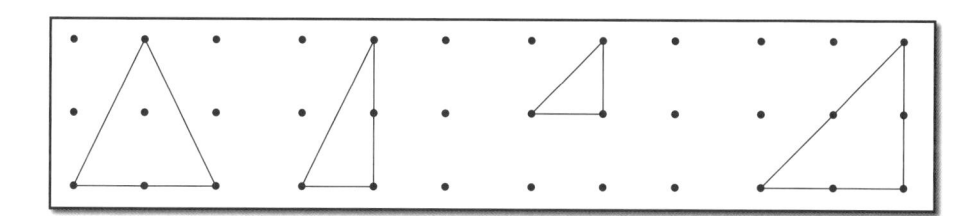

Label the acute angles 'A', the obtuse angles 'O' and the right angles 'R'. Write the name of each triangle underneath the drawing.

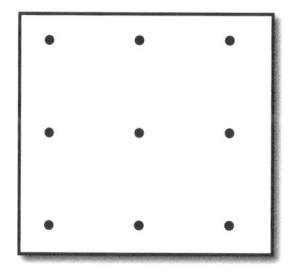

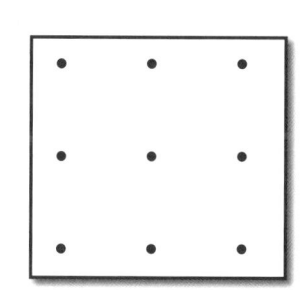

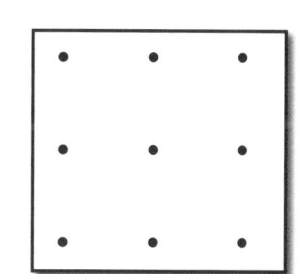

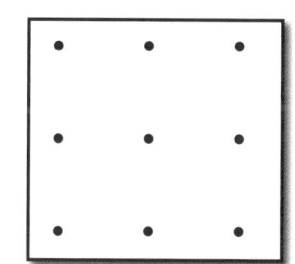

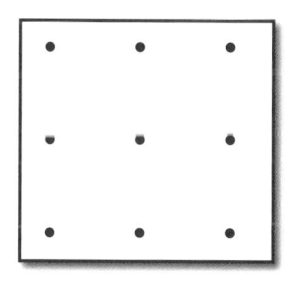

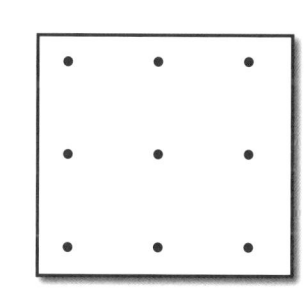

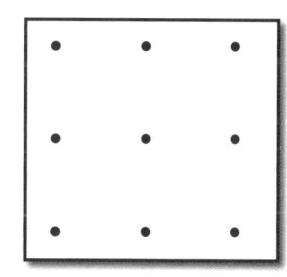

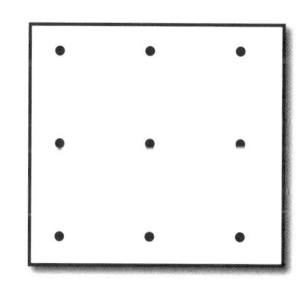

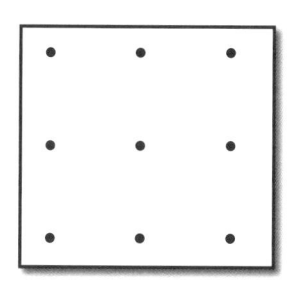

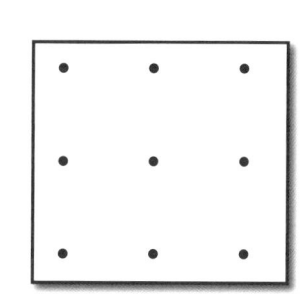

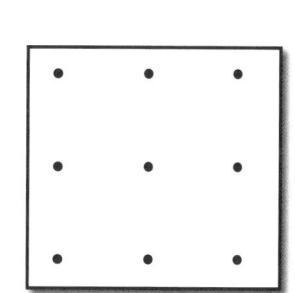

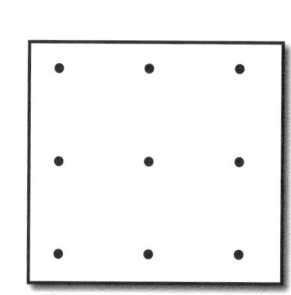

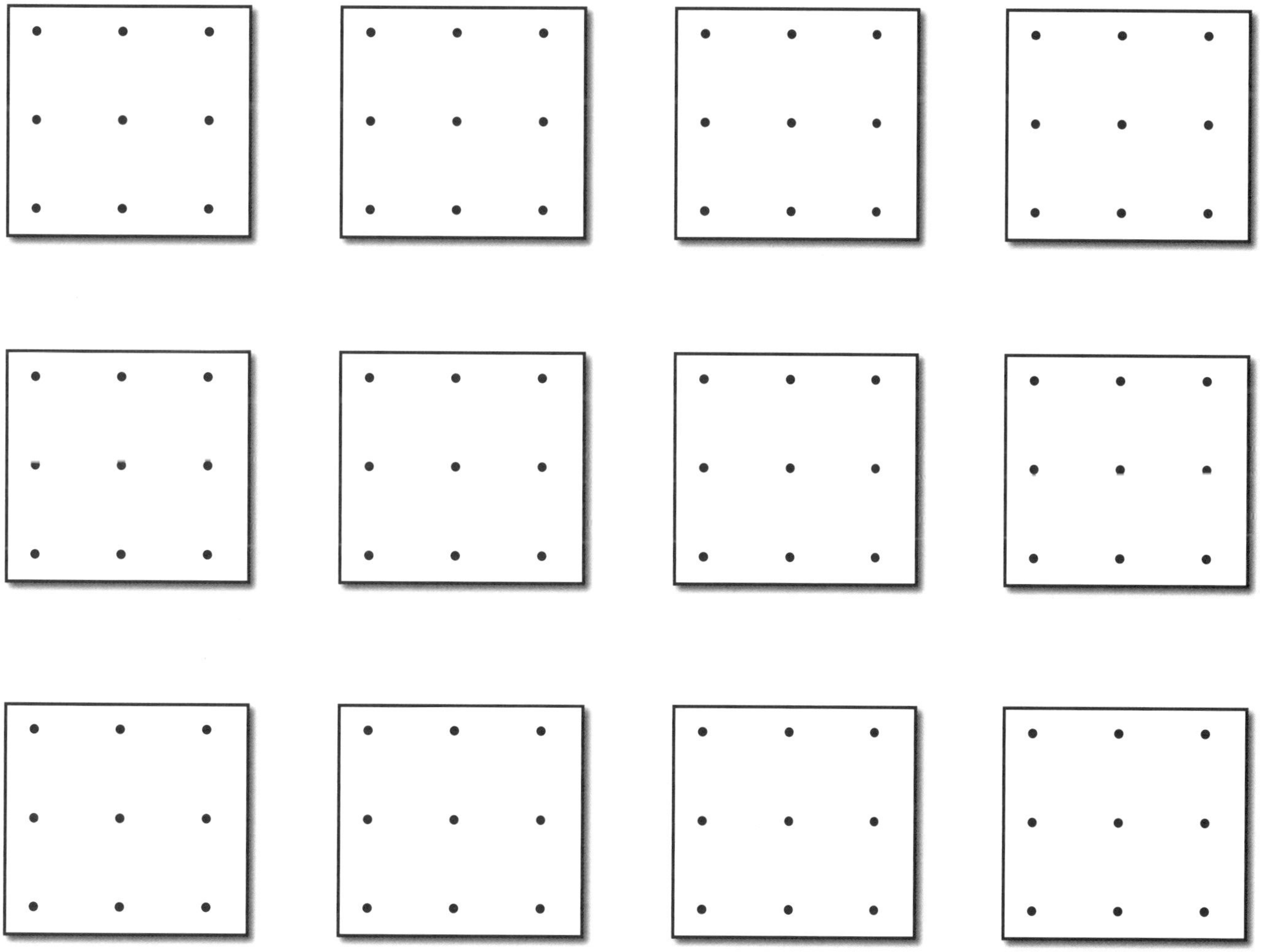

6 Review

Complete the table. Write whether each shape is 2D or 3D. Then write down three properties of each shape.

Shape	2D or 3D	Property 1	Property 2	Property 3
Scalene triangle				
Regular heptagon				
Hemisphere				
Tetrahedron				
Rhombus				
Dodecahedron				
Octahedron				
Kite				
Right-angled triangle				
Parallelogram				
Triangular prism				
Regular hexagon				

7 Position and Movement

Introduction

This unit develops students' ability to use mathematics to describe the position and movement of objects. We use coordinates on a numbered grid to describe position and we use transformations such as reflections, rotations and translations to describe how objects move around the grid.

Ways to help

Make copies of a coordinate grid like this one and use cut-out copies of shapes. Students can then physically move the shapes around the grid, giving the coordinates of the vertices and describing the movements. Encourage them to use the correct terminology when they do this.

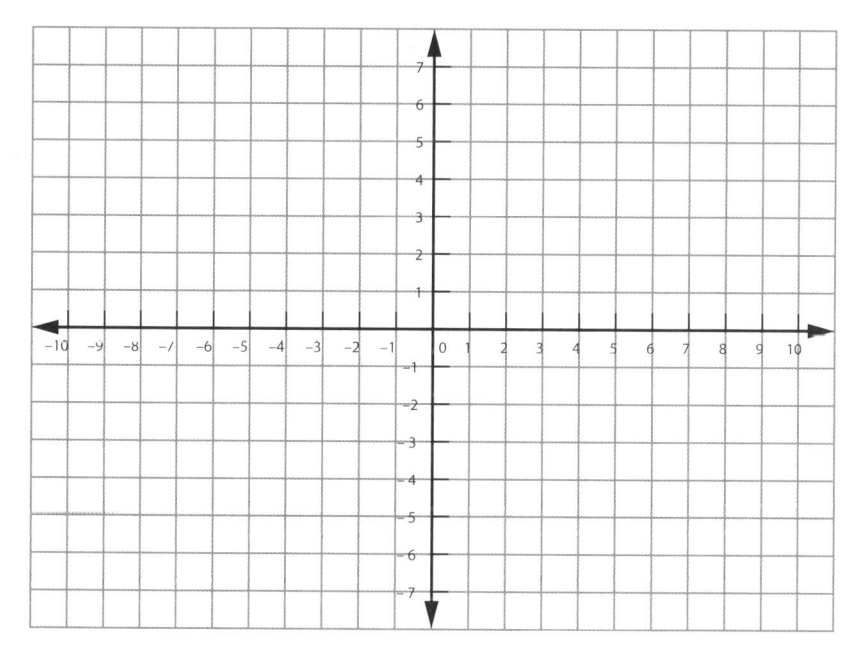

Key Words

origin; coordinates; first quadrant; second quadrant; third quadrant; fourth quadrant; symmetry; line of symmetry; axis of symmetry; mirror line; reflect; reflection; vertex; vertices; parallel; perpendicular; transformation; rotate; rotation; translate; translation; clockwise; anti-clockwise

7A Reading and plotting coordinates

Discover

Use this coordinate grid to draw a plan of a room in your house.
The origin should be in the centre of the room.

> Remember that the origin (0, 0) is where the two axes cross.

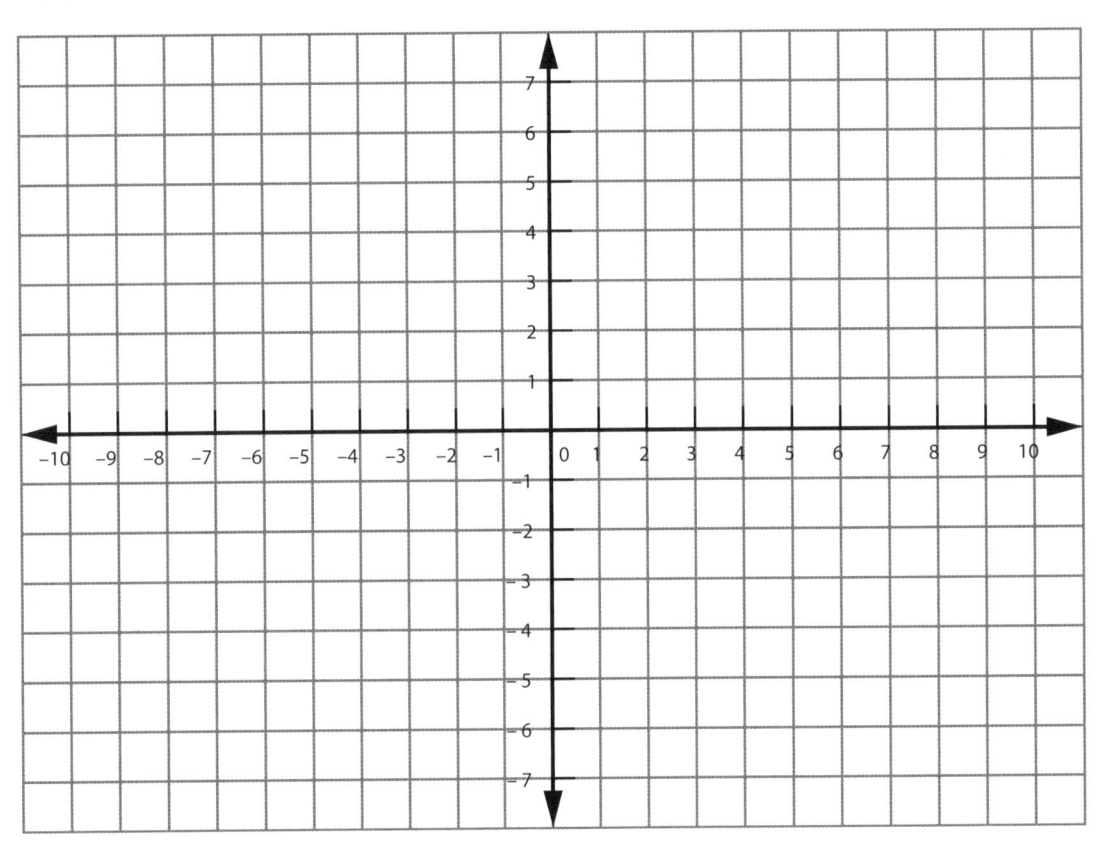

Write down the coordinates of four objects or pieces of furniture in the room.

a. The coordinates of _____ are (☐, ☐)

b. The coordinates of _____ are (☐, ☐)

c. The coordinates of _____ are (☐, ☐)

d. The coordinates of _____ are (☐, ☐)

7A Reading and plotting coordinates

Explore

You will need to work with a partner for this activity.
Draw an irregular hexagon somewhere in this grid.
Label the coordinates of the vertices.

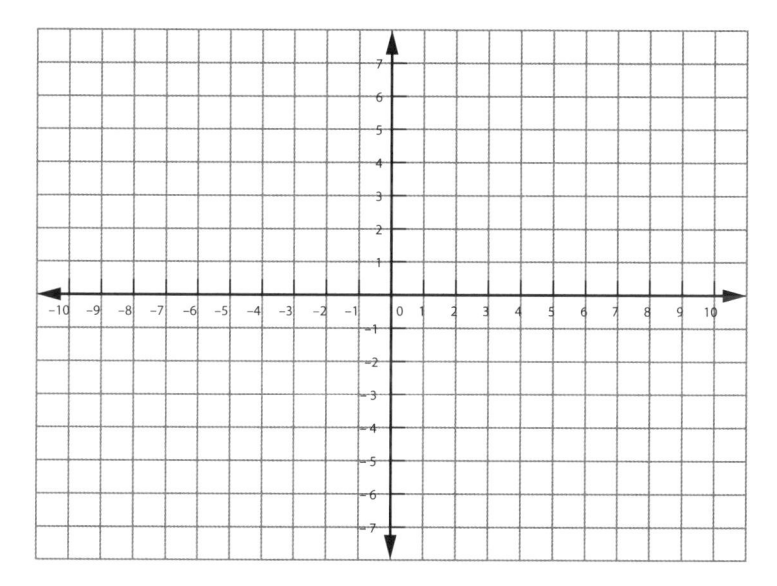

Cover up your shape with a piece of paper, then give this page to your partner.
Tell your partner the coordinates of the vertices of your hexagon.
Your partner should draw the hexagon in the grid below.

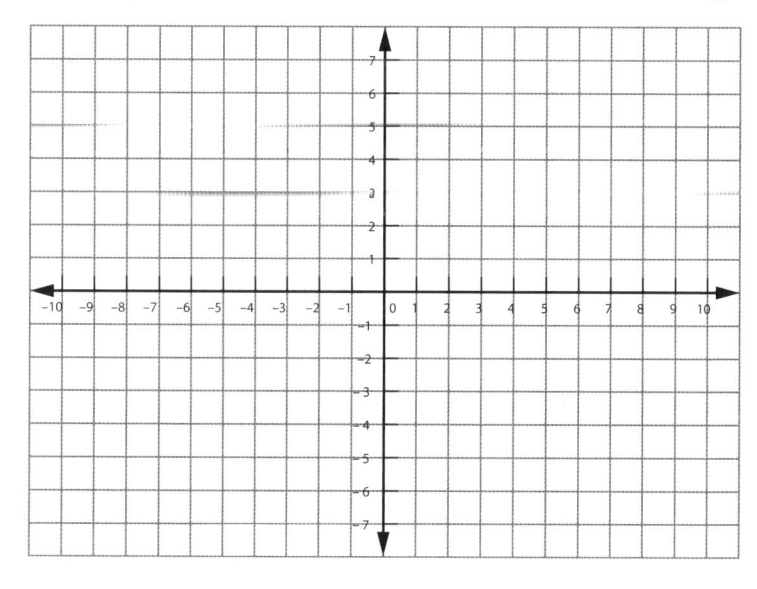

7B Reflections and rotations

Discover

Colour in the pattern so that it contains reflections, rotations and translations.

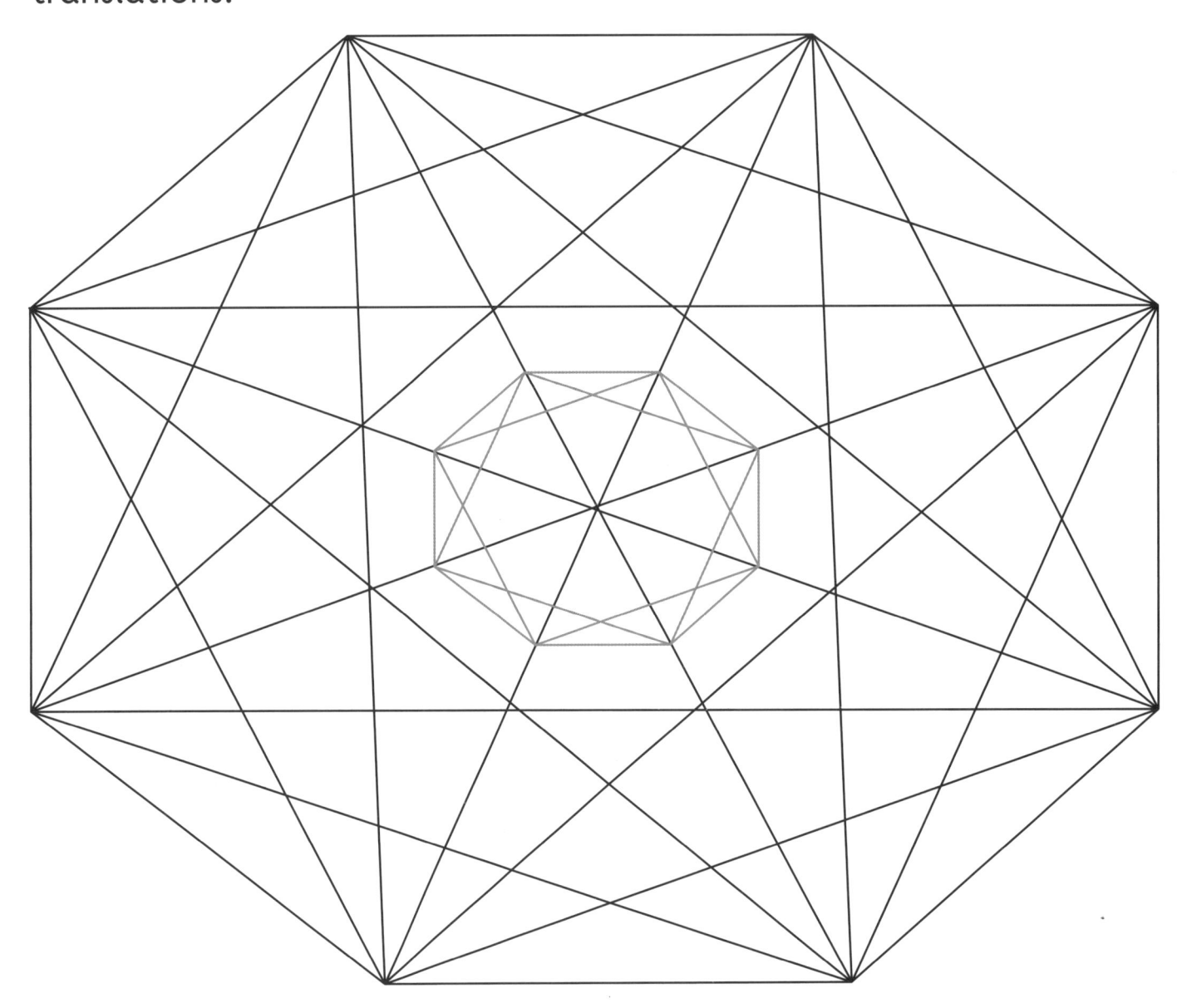

Label some of the reflections, rotations and translations that you can see.

7B Reflections and rotations

Explore

1. Draw a simple shape in the first quadrant on the grid below.
   One straight edge should be along the x-axis.
   Label the coordinates of the vertices.

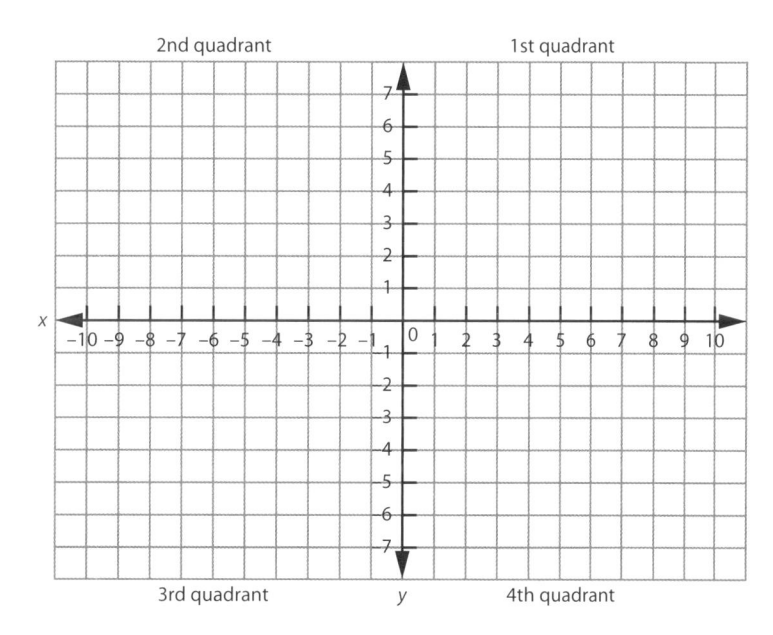

2. Reflect the shape in the x-axis and label the coordinates of the new vertices.

3. Reflect your shape from question 2 in the y-axis and label the coordinates of the new vertices.

4. Reflect your shape from question 3 in the x-axis.

5. What do you notice about the final reflection and the original shape?

_____

_____

7 Review

Draw your own pattern on the grid. Your pattern should include reflections, rotations and translations.

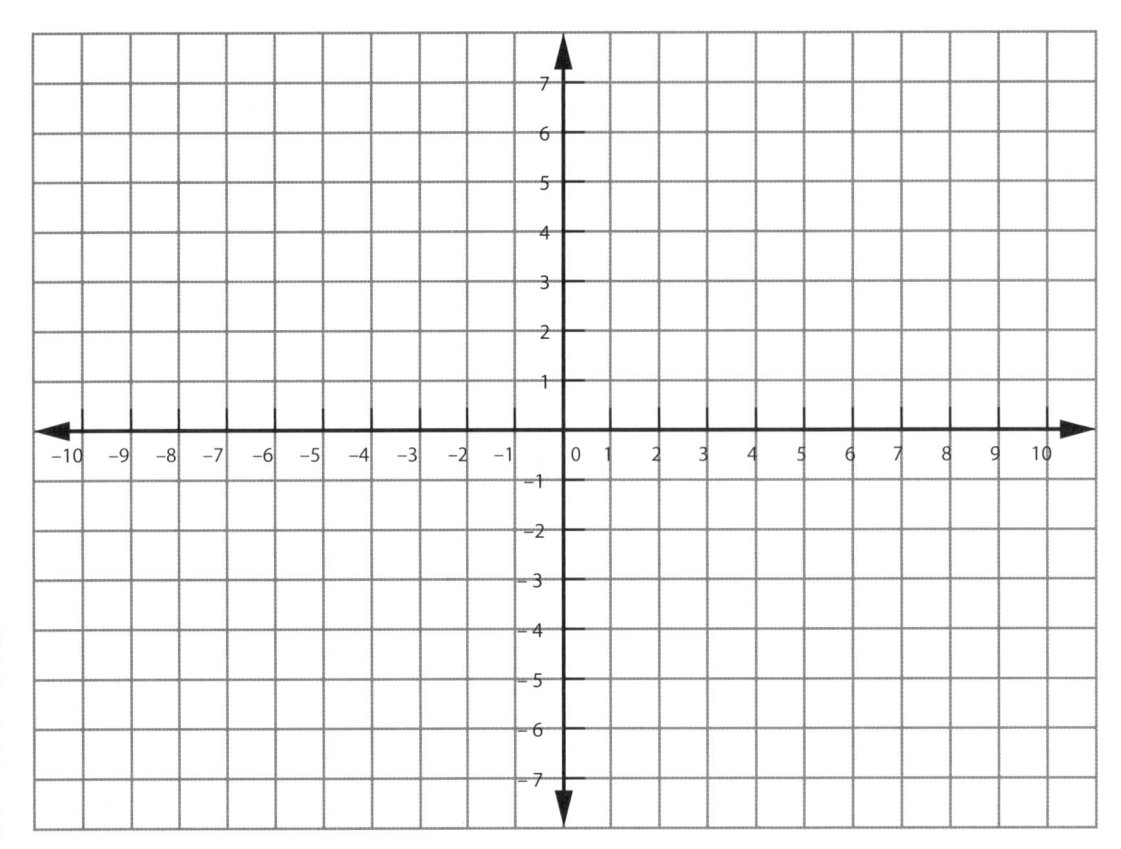

Write five sentences about your pattern. Include the words **reflection**, **rotation** and **translation**. Use coordinates to describe some parts of the pattern.

1. _____

2. _____

3. _____

4. _____

5. _____

8 Length, Mass and Capacity

Introduction

By Stage 6, students have a good understanding of how to use measures to compare length, mass and capacity. They will have an understanding of conservation, which means that an object has the same length whatever orientation it is in. They will also understand that smaller objects can have a greater mass than larger ones if they are made of denser materials. This unit develops students' ability to use their understanding of the decimal system to convert between units. They should remember that the prefix 'kilo' means 1000.

In this unit students are introduced to the idea that all measurements are approximations. For example, when measuring length we measure to the nearest millimetre, centimetre, metre or even kilometre, depending on what we are measuring and the required degree of accuracy. Students are also introduced to imperial units, which they will occasionally come across, particularly if they travel to other countries.

Ways to help

As in previous stages, the best way to help students learn how to measure is by physically measuring objects. Take any opportunity you can to measure length, mass and capacity. If you are making something that requires measuring, ask students to carry out the measurement for you. When cooking, ask students to carry out the measurements of mass and capacity that are required. Always estimate first – in this way students begin to develop a good sense of the size of the different units of measurement.

Key Words

metre; centimetre; millimetre; kilometre; mile; yard; feet; foot; inch; litre; millilitre; pint; gallon; tonne; kilogram; gram; pound; ounce; perimeter; area; square centimetre; cm^2

8A Selecting and using appropriate units of measure

Discover and explore

The table gives some different measurements but the units are missing!

Write in the correct metric unit for each measurement.

Then use the conversions in the box to convert the metric units to imperial units.

Conversions: metric to imperial

1 cm = 0.4 inch
1 m = 1.1 yards
1 km = 0.625 miles
10 g = 0.3 ounces
1 kg = 2.2 pounds
1 litre = 1.8 pints

All the conversions given here are approximate.

Object		Measurement in metric units	Conversion to imperial units
Height of the Shanghai Tower		632	
Length of a monitor lizard		160	
Mass of an elephant		2300	
Capacity of a large drinking glass		450	

8A Selecting and using appropriate units of measure

Discover and Explore

Object		Measurement in metric units	Conversion to imperial units
Mass of the heaviest human being		635	
Length of the River Nile		4180	
Capacity of a can of fizzy drink		330	
Mass of a small cat		2.9	
Distance from London to Dubai		5500	
Height of the smallest adult		55	

8B Converting units of measure

Discover

In this activity you will convert between different metric units of measurement.

1. First complete the conversion table.

2. Use the conversion table to work out these conversions.

Conversion table

1 km	=	_____ m
1 m	=	_____ cm
1 cm	=	_____ mm
1 l	=	_____ cl
1 l	=	_____ ml
1 cl	=	_____ ml
1 kg	=	_____ g
1 g	=	_____ mg

a. 8.7 kilometres in metres _____

b. 8.75 metres in centimetres _____

c. 89.5 centimetres in millimetres _____

d. 5.55 litres in centilitres and in millilitres _____

e. 8.45 centilitres in millilitres _____

f. 8.95 kilograms in grams _____

g. 9.65 grams in milligrams _____

h. 5500 metres in kilometres _____

i. 9850 millilitres in litres _____

j. 6750 grams in kilograms _____

8B Converting units of measure

Explore

You will need to convert between units of measure to solve these word problems.
You will need a calculator to help you and you may need an atlas and a stopwatch.

1. If you walk 1 million metres from your home where will you be? You can choose which direction to walk in.

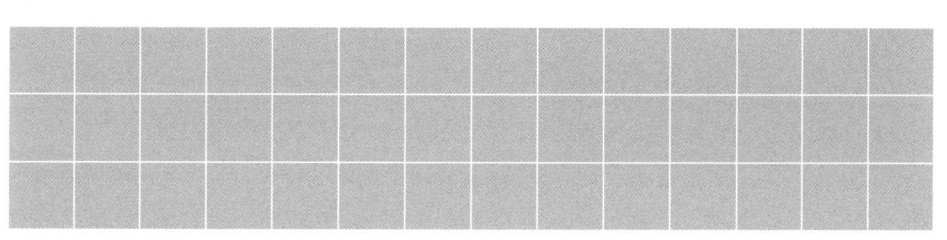

2. How old will you be when you have lived for another 1 000 000 minutes? Give your answer in years, months and days.

3. What is the mass of 1 000 000 grains of sand? (1 grain of sand = 0.05 mg)

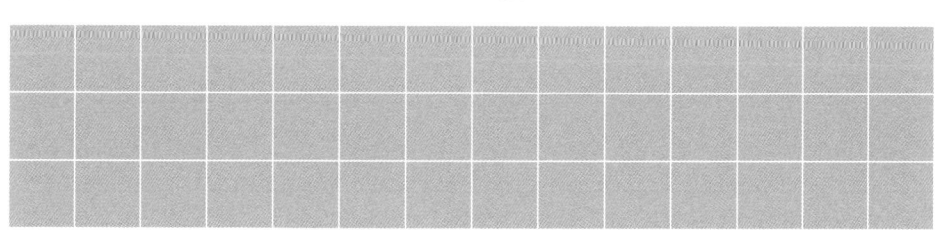

4. How much air do you breathe in a week? (1 breath = 500 ml)

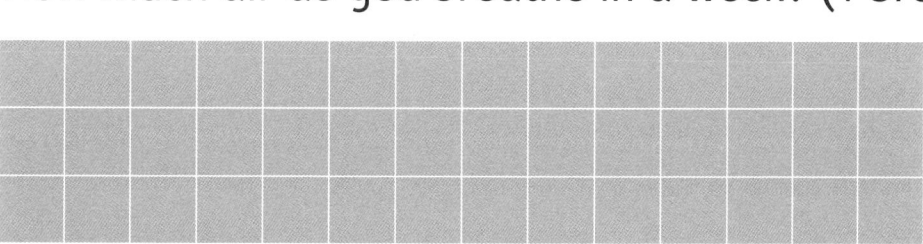

8C Using scales and measuring accurately

Discover and Explore

You will need a ruler, an elastic band and a piece of string for this activity.

You are going to design a pencil case to hold all your pens, pencils and other school equipment.

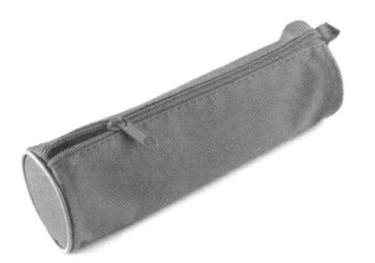

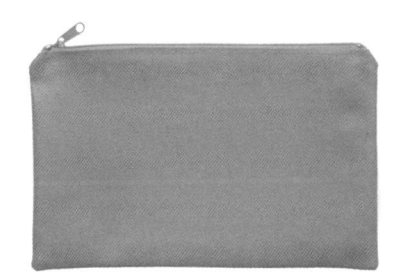

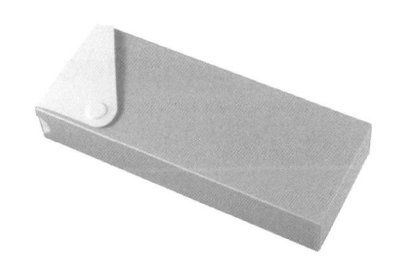

Answer the questions to help you design and make a pencil case that is just right for you.

1. What is the longest pen, pencil, ruler or other piece of equipment in your pencil case?

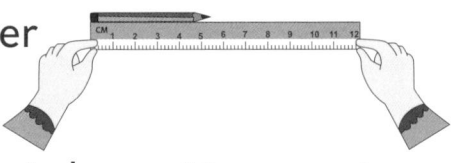

   List some of the items and their measurements here. Measure to the nearest millimetre.

   _____

   _____

   _____

   _____

   _____

8C Using scales and measuring accurately

Discover and Explore

2. Wrap an elastic band around all the items in your pencil case to hold them all together. Measure the length, width and depth that the pencil case will need to be to hold them all. Measure to the nearest $\frac{1}{2}$ cm.

   Length _____ Width _____ Depth _____

3. Use a piece of string to measure the distance around your pens, pencils and other equipment. Then use a ruler to measure the string. Measure to the nearest $\frac{1}{2}$ cm.
   Distance around the equipment _____

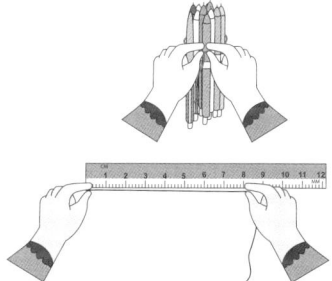

4. You now have all the dimensions you need for your pencil case.
   Decide what shape you want your pencil case to be. Do you want a cylinder, a cuboid, a triangular prism or another shape?

   _____

5. Draw the net for your pencil case. Label the sides with the dimensions.
   Your sketch of the net does not have to be to scale.

8 Review

1. A bottle of water holds 330 ml. There are 30 people in my class, who drink a bottle each. How much water do they drink altogether? Give the answer in litres.

2. You are packing for a holiday. The weight limit for your suitcase is 18 kg. You want to pack a laptop that weighs 4.5 kg, five books that weigh 300 g each and clothes that weigh 9.5 kg. How much does everything weigh in total?

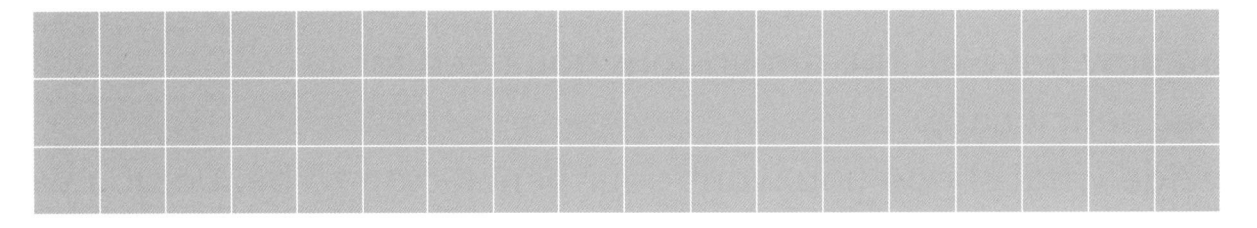

   How much of your weight limit do you have left?

3. I am going to post a package. There are five smaller parcels in the package. They weigh 250 g; 350 g; 1.65 kg; 1500 g and 1.25 kg. The postage is $1.50 for every 500 g. How much will it cost to post the parcel?

9 Time

Introduction

This unit continues to develop students' skills in telling the time. They tell the time to the nearest minute and use analogue and digital clocks. They extend their skills in reading timetables and calendars and carry out calculations involving converting between units of time. They are introduced to the concept of time zones and learn how to calculate the time in places in different time zones around the world.

Ways to help

Telling the time is something that we do all the time. We use our skills in telling the time every day in order to plan our day and make sure that we are not late for appointments. Whenever you notice yourself glancing at the clock or using a timetable to see what is on television or to plan a journey, ask students to do this task with you.

It would be useful to have both an analogue clock and a digital clock in your home so that students become used to both ways of representing the time.

You can explore different time zones using the Internet. There are several sites that show clocks telling the time in different cities around the world. If any members of your family live in different countries, ask students to work out what the time is where they are. For example, at bedtime ask: 'What time is it where [name of family member] lives? What do you think they are doing now?'

Key Words

millennium; century; decade; year; month; fortnight; week; day; hour; second; millisecond; leap year; a.m.; p.m.; timetable; arrive; depart; international time zones

9A Converting between units of time

Discover

For each question you are given a starting time and a duration.

Write the starting time on the digital clock and draw the end time on the analogue clock.

The first one is done for you.

1. Six minutes to nine in the morning
   2 hours 13 minutes later

   08:54

2. Eighteen minutes past eleven in the morning
   3 hours 28 minutes later

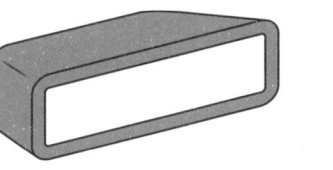

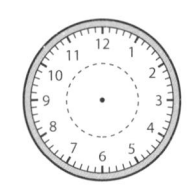

3. Twenty-five minutes to five in the afternoon
   5 hours 35 minutes later

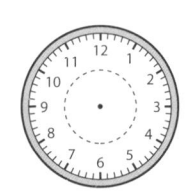

4. Seventeen minutes past ten in the evening
   4 hours 28 minutes later

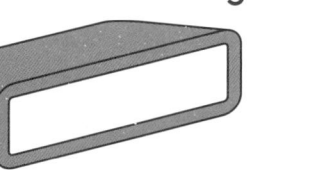

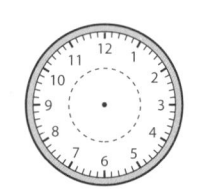

5. Five minutes to midnight
   7 hours 39 minutes later

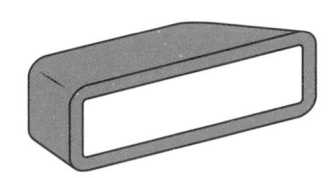

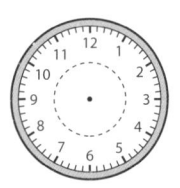

9A Converting between units of time

Explore

You could use the table that you completed on page 152 of the Student Workbook to help you answer these questions.

1. I was born on 8 June 1959. How many days old am I?

_____ days

2. African elephants live for an average of 57 years in the wild and only 17 years in captivity. How many weeks longer do elephants in the wild live than elephants in captivity?

_____ weeks

3. Al-Khwārizmī wrote a famous mathematics book in the year 830. How many decades ago did he write the book?

_____ decades

4. What date is it one million days after the start of this millennium?

_____

5. How many milliseconds are you at school each day?

_____ milliseconds

9B Using the 24-hour clock and timetables

Discover

Use the grid to plan your ideal timetable for a week at school.
Write the days that you attend school in the top row.
Write the times of the lessons and breaks in the left-hand column.
Use the 24-hour clock. You can decide your own start and finish times.

Write five facts about the timetable.

1. _____

2. _____

3. _____

4. _____

5. _____

9B Using the 24-hour clock and timetables

Explore

Design a flight timetable from Dubai airport for one day (24 hours), using these flight times and rules. Write your timetable below.

Flight times from Dubai airport:

Abu Dhabi	30 minutes	Doha	1 hour 20 minutes
Abujah	7 hours 20 minutes	Islamabad	3 hours 3 hours 5 minutes
Amman	3 hours 10 minutes	Jeddah	3 hours 2 minutes
Bangkok	6 hours 30 minutes	Manchester	7 hours 35 minutes
Cairo	3 hours 45 minutes	Moscow	5 hours 20 minutes

Rules:

- There must be one flight to each destination.
- The first flight cannot leave before 06:30 and the last flight must leave before 23:15.
- There must be at least half an hour between departure times.
- No flights are allowed to leave between 11 am and 1 pm.
- The Doha flight must arrive in Doha before 17.00.
- Spread departure times as evenly as possible throughout the day.

Destination	Departure time	Arrival time (Dubai time)

9C Calculating time intervals including time zones

Discover

These clocks show you the time differences between some cities around the world.

NEW YORK 04:15
LONDON 09:15
RIYADH 11:15
DUBAI 12:15
BANGKOK 15:15
SYDNEY 18:15

Use this information to calculate the following times. The first one is done for you.

1. What time is it in Riyadh when it is 13:25 in London?

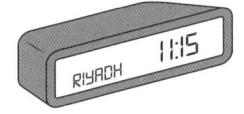

   15.25

2. What time is it in Sydney when it is 07:30 in Dubai?

   _____

3. What time is it in Bangkok when it is 15:45 in Riyadh?

   _____

4. What time is it in London when it is 17:15 in Dubai?

   _____

5. What time is it in New York when it is 09:25 in Bangkok?

   _____

6. What time is it in Dubai when it is 23:30 in Sydney?

   _____

7. What time is it in London when it is midnight in Dubai?

   _____

8. What time is it in Sydney when it is noon in New York?

   _____

9C Calculating time intervals including time zones

Explore

The timetable shows a list of flight departures from Dubai airport.

Use this list of time differences to help you complete the timetable. The first one is done for you.

Time differences compared with Dubai:

Sydney, Australia	+ 6 hours
Dhaka, Bangladesh	+ 2 hours
Tokyo, Japan	+ 5 hours
Nairobi, Kenya	−1 hour
Phnom Penh, Cambodia	+ 3 hours
Kuala Lumpur, Malaysia	+ 4 hours
Toronto, Canada	−8 hours
Anchorage, Alaska	−12 hours
London, United Kingdom	−3 hours

Hint: You need to add the flight duration to the departure time and then add or subtract the time difference.

Destination	Departure time	Flight duration	Arrival (local time)
Sydney	09:15	12 hours 30 minutes	03:45 (the next day)
Dhaka	10:25	5 hours 15 minutes	
Tokyo	10:45	9 hours 10 minutes	
Nairobi	13:35	5 hours 25 minutes	
Phnom Penh	15:50	6 hours 45 minutes	
Kuala Lumpur	19:25	7 hours 15 minutes	
Toronto	20:25	15 hours 45 minutes	
Anchorage	22:10	17 hours 45 minutes	
London	22:15	6 hours 50 minutes	

9 Review

Use your knowledge of time to solve these word problems.

1. The first lesson in the afternoon starts at 13:15. There are two lessons of 50 minutes and a 15-minute break. Then there is 10 minutes in the home room before the end of school. What time does school finish?

2. Your uncle is working in a country where the time is 7 hours behind the time in your country. You want to telephone him when the time is 13:30 in his country. What time should you telephone in your time zone?

3. A train leaves the station at 17:38. The journey takes 2 hours 36 minutes. What time does it arrive at its destination?

4. How many seconds are there in one week?

10 Area and Perimeter

Introduction

Students have learned how to calculate the area (amount of space covered by a shape) and the perimeter (the distance around the edge of a shape) of shapes in previous stages. In this unit they learn that we use standard measures for these measurements. Perimeter is measured in units of length (cm, m, km, etc.) and area is measured in square units (cm^2, m^2, km^2, etc.). This means that we calculate an area by finding out how many unit squares would fit inside the shape.

Ways to help

Sometimes students confuse area and perimeter. You can help by looking at shapes and pointing out which is the perimeter and which is the area. You can use small square objects such as tiles or squares cut from paper to make different shapes. First make a shape from some squares and help students to work out the area and perimeter. Then rearrange the squares to make another shape and work out the new area and perimeter. As you have used the same number of squares, the area will be the same but the perimeter may be different. Here is an example.

This shape has perimeter 10 units and area 4 square units.

This shape has perimeter 8 units and area 4 square units

> **Key Words**
> length; width; height; depth; breadth; edge; perimeter; circumference; millimetre; centimetre; metre; area; surface; surface area; square centimetre; square metre; cm^2; m^2

10A Area and perimeter of rectilinear shapes

Discover

You are going to design a school playground.

You have enough money to buy 40 square metres of a special surface that will make the playground safe to play in.

Draw three different designs for the playground that each have an area of 40 square metres.

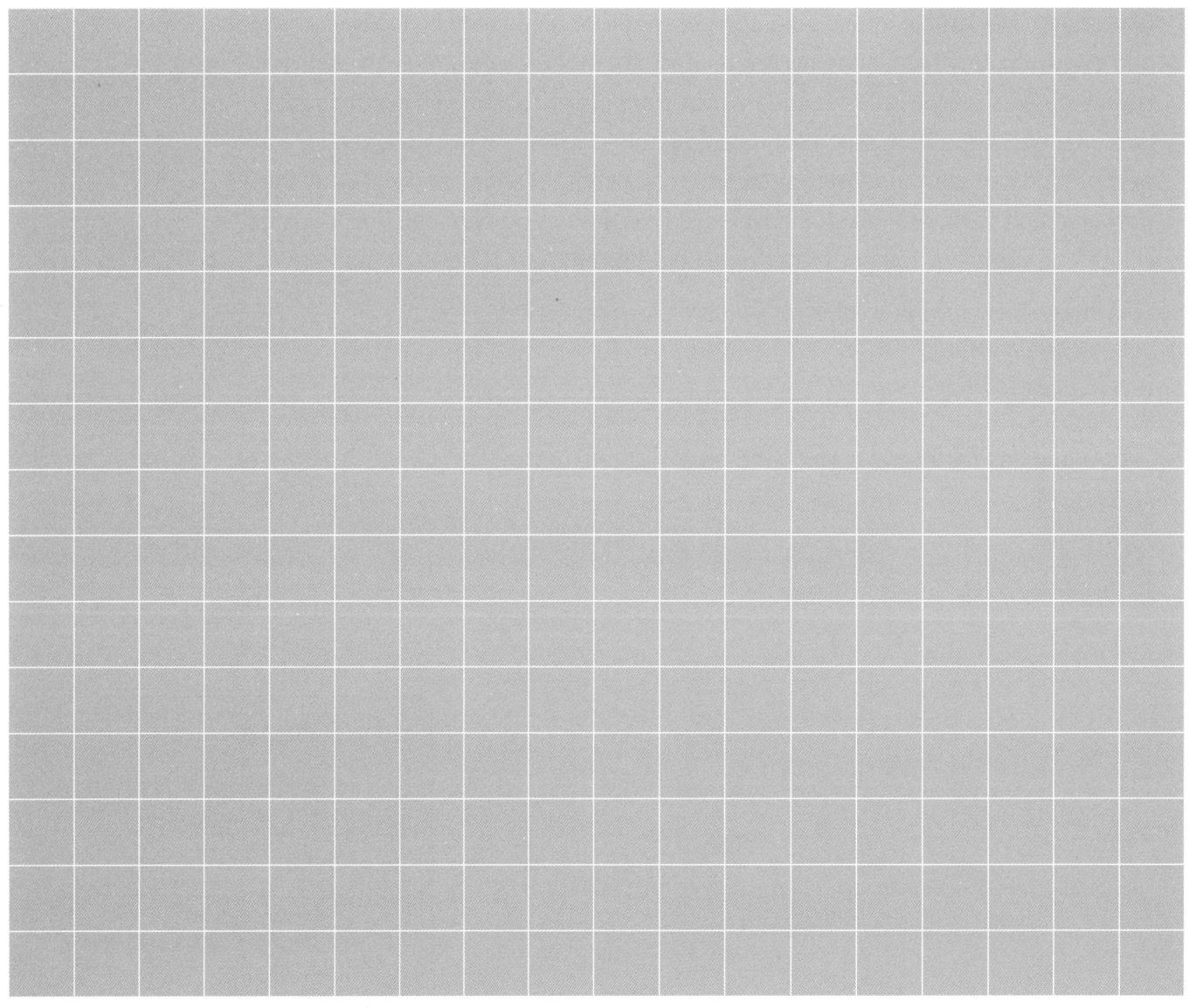

10A Area and perimeter of rectilinear shapes

Explore

Draw rectilinear shapes that have the following properties on the squared paper below. Label each shape with its letter.

A A rectilinear shape with a perimeter of 20 cm

B A rectilinear shape with an area of 18 cm²

C A rectilinear shape with a perimeter of 28 cm

D A rectilinear shape with an area of 36 cm²

E A rectilinear shape with an area of 40 cm²

F A rectilinear shape with a perimeter of 32 cm

10B Estimating areas of irregular shapes by counting squares

Discover

Find three leaves of different sizes.

Place each leaf on the grid below and draw around each one.

Estimate the area of each leaf by counting squares. Write the area inside each leaf.

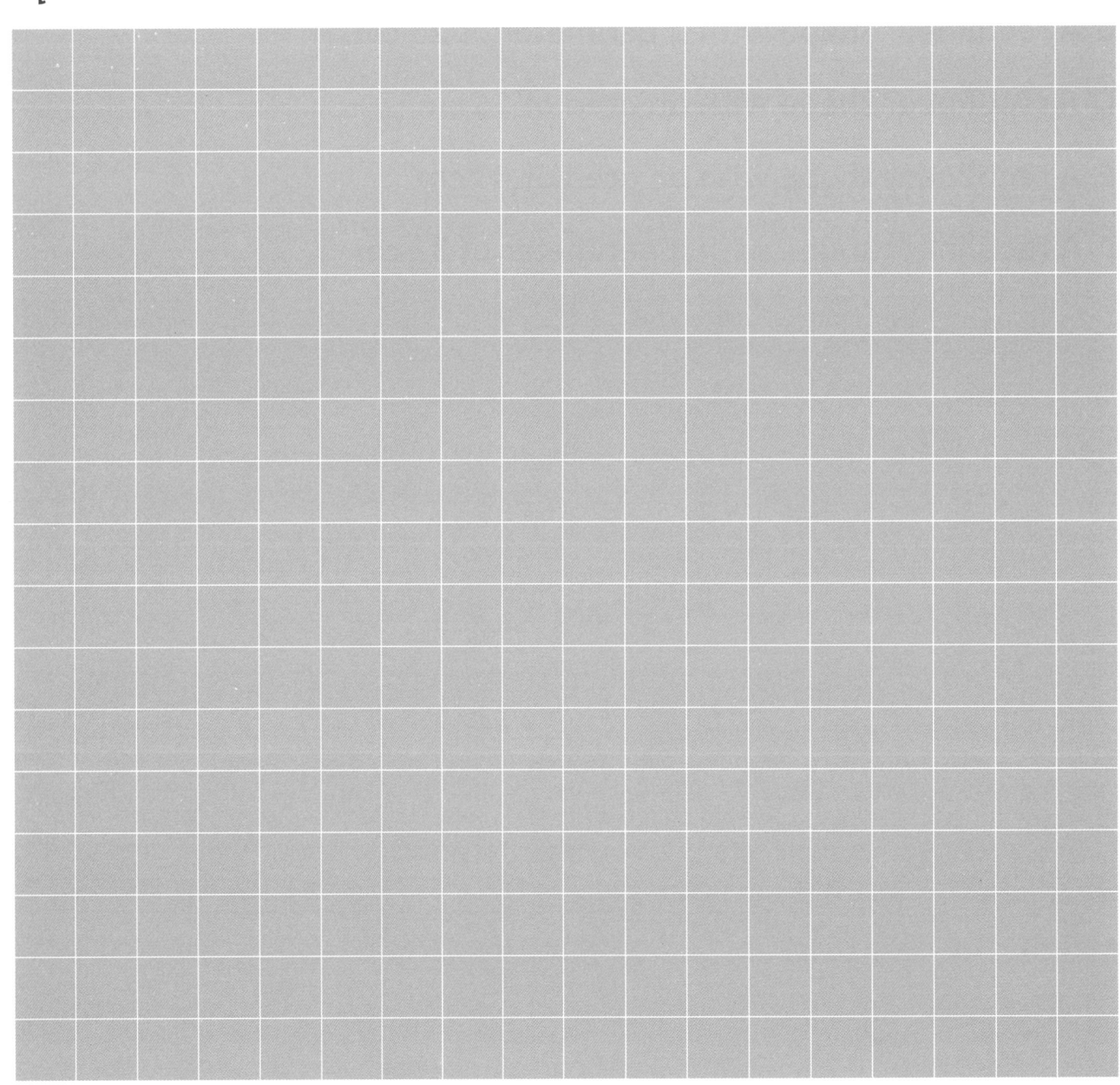

10B Estimating areas of irregular shapes by counting squares

Explore

Look again at the leaves you drew on page 88 to help you get an idea of the areas of leaves of different sizes.

Then sketch leaves with the following areas on the squared paper below.

Sketch each leaf first and then count the squares to check the area. Label each shape with its letter.

A 38 cm²

B 46 cm²

C 19 cm²

10C Calculating areas and perimeters of compound shapes

Discover

Calculate the areas and perimeters of these shapes. Remember to give the unit of measurement in your answers. The shapes are drawn on 1 cm squared paper.

Shape	Perimeter	Area
A		
B		
C		
D		

10C Calculating areas and perimeters of compound shapes

Explore

Find a cardboard box at home or ask for one at a local shop. Sketch the net of the box on the grid below. It does not have to be to scale.

Write down the surface area (total area) and the perimeter of the net.

Surface area: _____

Perimeter: _____

10 Review

Draw a plan of a room in your house on the squared paper below. Add plan views of the furniture in the room, using straight lines for the edges of the furniture.

Your plan does not have to be to scale but the dimensions should be realistic.

Work out the total area of floor space that is left when you have drawn all the furniture on the plan.

Work out the total of all the perimeters of the pieces of furniture added together.

Area of floor space with no furniture:

Total of all the perimeters of the furniture:

11 Handling Data

Introduction

In this final unit students develop their understanding of handling data. They decide on the best type of graph or chart to show different types of data and work out the scales to use. Students also develop their understanding of probability. New ideas in this unit include measures of the 'average' and 'spread' of a set of data such as the mean, median, mode and range.

To calculate the **mean**, add together all the numbers in the set and divide by the number in the set. To find the mean of 14, 15, 16, 18, 18, 18, 21, 22, add all the numbers together and divide by 8:

$$\frac{14 + 15 + 16 + 18 + 18 + 18 + 21 + 22}{8} = \frac{142}{8} = 17.75$$

The **median** is the middle number when the set is arranged in ascending order. For example, the set 14, 15, 16, 18, 18, 18, 21, 22 has an even number of items, so there are two numbers in the middle – they are both 18 so the median is 18.

In the set 14, 15, 16, 18, 19, 20, 21, 22 we take the number halfway between the two middle numbers; the number halfway between 18 and 19 is 18.5.

It is easier to find the median of sets of data with an odd number of items because there is only one middle number. The median of 15, 16, 18, 19, 20, 21, 22 is 19.

The **mode** is the number which occurs most frequently or most often in a set. The mode of the set 14, 15, 16, 18, 18, 18, 21, 22 is 18.

The **range** shows the spread of the data. It is the difference between the largest and smallest values. The range of the set 14, 15, 16, 18, 18, 18, 21, 22 is 22 − 14 = 8.

Ways to help

Look for graphs and charts presenting data, for example on the Internet and in newspapers. Talk to students about how this data might have been collected. Challenge them to work out the mean, median, mode and/or range of the data where appropriate.

Key Words

mean; median; average; mode; range; bar line graph; line graph; frequency table; axis; axes; scale; interval; most popular; most common; least popular; least common; survey; questionnaire; data; continuous data; discrete data; event; outcome; probability; certain; impossible; equal chance; fair; unfair; biased; equally likely outcomes

11A Handling data

Discover and Explore

In the first column of the table, write the names of your home country and two other countries.

Use the Internet to find the mean temperatures each month in all three countries. Use what you find out to complete the table.

	Jan	Feb	Mar	Apr	May	Jun	Jul	Aug	Sep	Oct	Nov	Dec
My country ___												
Country A ___												
Country B ___												

Write down three facts about this data. Include a fact about the range of temperatures and a fact about the mode temperature.

1. _____

2. _____

3. _____

Choose and draw a suitable chart or a graph to present this data.

11B Probability

Discover and Explore

Complete the table to find the probabilities of these events. The first one is done for you.

Chosen outcome	Equally likely outcomes	Number of possible required outcomes	Probability fraction
Picking a Jack, Queen or King from a pack of cards	There are 52 cards to pick from	12 (There are 3 picture cards in each suit and there are 4 suits in the pack.)	$\frac{12}{52} = \frac{3}{13}$
Rolling an odd number on a 1–6 dice			
Rolling a total of 9 on two dice			
Rolling an even number on a 1–10 dice